THE BAOFENG RADIO REVOLUTION

THE BEGINNER GUERRILLA'S GUIDE TO BREAK
THROUGH THE COMPLEXITY, SECURE
COMMUNICATIONS, AND PREPARE FOR DISASTER
WITH PREPPER TACTICS

MORSE CODE PUBLISHING

CONTENTS

INTRODUCTION

BAOFENG SURVIVAL

When Alden Jones packed his backpack for a solo hike along Vermont's picturesque Long Trail, his expectations were set on breathtaking vistas and serene solitude amidst nature's embrace. Little did he know, this journey would thrust him into a harrowing ordeal, where a simple device – a Baofeng radio – would become his lifeline (Tate, 2021).

Stricken by a sudden diabetic emergency, Alden's situation turned critical. Miles away from the nearest town, deep in the wilderness, his cell phone was as good as a brick. The sun dipped below the horizon, bringing the chilling prospect of a night spent helpless in the wild. But Alden wasn't alone in his plight as a fellow hiker came to his aid. Alden's foresight in bringing along his Baofeng handheld radio, a modest yet powerful tool, was about to pay off.

Activating his radio, Alden transmitted a desperate mayday call across a local amateur radio frequency. Miraculously, his call was answered. Two dedicated ham radio enthusiasts, equipped with their Baofeng radios, picked up his distress signal. They mobilized swiftly, coordinating a helicopter rescue that would whisk Alden to safety. This incident wasn't just a testament to Alden's survival instinct; it underscored the critical role a small, battery-powered Baofeng radio could play in emergencies, even in the most secluded corners of our planet.

Inspired by Alden's real-life survival tale, I embarked on my own journey into the world of radio and preparedness. As a novice, the realm of radio communication was daunting, filled with technical jargon and complex operations. Yet, I was convinced of its invaluable role in disaster scenarios where conventional communication methods fall short. This book is my foray into understanding Baofeng radios, focusing on the popular UV-5R model.

My aim is straightforward: to introduce beginners to the fascinating world of Baofeng radios. Through my experiences, I've gained insights into selecting, setting up, and troubleshooting these devices. This book is my conduit for sharing this knowledge, simplifying the technical jargon of ham radios, and empowering anyone to operate a Baofeng radio confidently. There is no better feeling of safety in a disaster than knowing you have functioning communications.

THE BAOFENG REVOLUTION

In these pages, you will find a comprehensive guide to Baofeng radios, tailored specifically for beginners. While I discuss the Baofeng product line broadly, I use the UV-5R model as a practical example to illustrate setup instructions and operational tips. My goal is to equip you with the knowledge to tune into your first frequency – your local NOAA Weather Radio station – and, from there, embark on a thrilling journey into the world of amateur radio.

This guide covers various topics, from choosing the right model, initial setup, understanding ham radio lingo, and programming your device to enhancing signal clarity, understanding legal aspects, exploring advanced features, and applying your knowledge in real-world scenarios. Furthermore, I delve into emergency communication strategies, routine maintenance, and effective troubleshooting techniques. By the end of this book, you'll possess the knowledge to use your Baofeng radio independently and develop a deep appreciation for its capabilities.

While numerous guides cater to seasoned radio enthusiasts, this book fills a critical gap, offering practical and accessible guidance for newcomers. My mission is to empower you, the Baofeng beginner, with the competence and confidence to handle your radio proficiently, regardless of your prior experience. I aspire to ignite in you the same passion for Baofeng radios that now fuels my enthusiasm.

So, I invite you to join me in uncovering the tactical potential of the little Baofeng radio that could. This journey is not just about mastering a piece of technology; it's about discovering how this technology can simplify, secure, and even save our lives in times of need. Whether standing before an array of radios in a store, feeling overwhelmed, or simply curious about communication technology, this book is your beacon.

Baofeng radios are the ideal entry point for beginners venturing into the realm of communication technology. Comparable to a guide for novice mountain climbers, these radios ensure safety, simplicity, and enjoyment as you navigate the peaks and valleys of radio communication. Their appeal lies in their affordability, performance, and versatility. The real question isn't why you should choose Baofeng radios, but why wouldn't you?

As we progress through this book, we will demystify the jargon, untangling the complex web of frequencies, channels, and signals. Fear not; the world of amateur radio is far less intimidating than it appears. Remember, every expert was once a beginner. I penned this book initially to aid my learning journey and soon realized that in doing so, I could help others embark on their own path.

The Baofeng radio is more than a mere gadget; it's a steadfast companion. It stands by you in emergencies, offering a crucial connection when other means falter. It accompanies you in your leisure, linking you to communities across the globe. This book sets the stage for your adventure into the Baofeng universe. We'll guide you through setting up your first radio, maintaining it, programming it, and resolving basic operational issues. You don't need to be a tech whiz; you only require curiosity and a zeal for learning.

As you turn each page, prepare to unearth a new passion, hobby, and skill set. Welcome to the Baofeng Radio Revolution – your journey begins now.

Baofeng UV-5R

Frequency Offset Direction For Accessing Repeaters

Reverse Function Activated

Wide Band Selected

Keypad Lock Function Activated

Function "VOX" Enabled

Signal Strength Level

CTCSS Activated

DCS Activated

Operating Frequency

Operation Frequency

Operation Channel

Battery Level Indicator

DISPLAY

LCD

VFO/MR

SP&MIC

Speaker

BAND Key

Battery Pack

Antenna

Knob
ON/OFF, Volume

Flashlight

Battery Remove
Button

SK-side Key 1/CALL
(radio. alarm)

PTT key

LED Indicator

MONI Key

A/B Key

Keypad

Battery Contacts

1

THE BAOFENG BLUEPRINT - AN INTRODUCTION TO HANDHELD RADIO & CHOOSING A BAOFENG MODEL

In the world of emergency preparedness, the ability to communicate effectively can be the difference between safety and peril. The journey into two-way radio communication is not just a hobby; it's a critical aspect of ensuring resilience in the face of emergencies. This chapter aims to guide you through the diverse landscape of Baofeng two-way radio models, which are renowned for their reliability and affordability. Choosing the proper radio is a strategic decision, pivotal in establishing dependable communication lines during crises when traditional networks might fail.

Understanding the nuances of Baofeng radios is crucial, whether you're coordinating with a community network during a natural disaster, staying in touch with your group during off-grid adventures, or simply ensuring a line of communication when other systems are down. This chapter will help you navigate the options available, aiding you in selecting a device that aligns with your specific preparedness goals and scenarios. It will also ensure you're well-equipped to communicate effectively in any situation.

THE VERY BASICS OF RADIO

In choosing your ideal Baofeng radio model, it helps to have a fundamental understanding of radio communication. In simple terms, two-way radios allow the wireless transmission of radio waves between a transmitter and a receiver (Ricky, 2022). The transmitter converts electrical signals into radio waves, which are picked up by the receiver's antenna. The receiver then converts the radio waves into electrical signals (*Problem With Baofeng BF-888 Radios,* n.d.).

Understanding radio waves is also crucial. Radios mainly use these types of waves (W0SJF, n.d.):

- VHF or very high-frequency waves (136–174 MHz range): Excellent outdoors over open terrain.
- UHF or ultra-high-frequency waves (400–520 MHz range): Great for penetrating obstacles like buildings and forests.

Here's my trick for remembering these differences: For VHF, focus on the 'V' for very long distances. For UHF, the 'U' reminds me of Urban, so I remember it's better at penetrating through buildings.

Baofeng radios can transmit on both VHF and UHF frequencies, a feature known as dual-band operation. This versatility is part of what makes them so popular.

The Radio Spectrum

To understand how radios work, one must have a basic grasp of the radio spectrum. This refers to the full range of radio frequencies used for wireless communications (*Baofeng Cheat Sheet,* n.d.).

Why is this information important? Understanding frequency fundamentals lets you choose the right radio equipment for your intended application.

The spectrum is divided into different bands based on frequency and wavelength, such as AM, FM, VHF, UHF, etc. Each band has unique propagation characteristics (the way radio waves travel through the air).

VHF refers to very high-frequency signals in the 136–174 MHz range. The wavelength ranges from 1–2 meters. VHF provides excellent propagation outdoors for services like TV, FM radio, and two-way land mobile radios.

UHF refers to ultra-high frequency signals from 400–520 MHz with wavelengths around 0.5 meters, or you may see it called 50-70 centimeters. Major uses include cell phones, WiFi, Bluetooth, two-way radios, satellite communication, and microwave ovens.

It is also important to note that the full VHF and UHF ranges are larger than the practical range used in radio service. The full VHF range is 30–300 MHz, and the full UHF range is 300–3000 MHz (3 GHz).

Higher frequency bands allow faster data transmission rates but have shorter transmission ranges. Lower frequencies travel further but with less bandwidth.

Scan the QR code or click on the link to receive 'The Baofeng Radio Revolution Command Packet,' which includes an overview of the full radio spectrum.

www.MorseCodePublishing.com/Command

Next, let's get to know more about the primary radio services.

PRIMARY RADIO SERVICES

Two-way radios utilize radio services, such as FRS, GMRS, MURS, and Ham radio. These differ based on licensing requirements, frequencies, transmission power, and a few other differences (Best Baofeng Radios, 2022).

FRS—Family Radio Service

The Family Radio Service (FRS) is a set of frequencies the FCC has designated for personal communication, much like an express lane reserved for family vehicles. These channels require no license and can be used for both routine and emergency communication. Key attributes include (Family Radio Service, 2011):

- FRS is a short-distance two-way radio service that anyone in the US can use without a license.
- Operates on 22 UHF channels within 462 and 467 MHz ranges.
- These channels are in the UHF band (great for close urban areas or dense forests).
- Limited power output requirements max out at 0.5 watts ERP (effective radiated power).
- Because of the UHF and limited power, the unobstructed range is generally under 2 miles.
- These are ideal for activities like hiking, skiing, and camping, where groups want to coordinate, and no licenses are needed.

FRS radios are the "walkie-talkies" you can purchase off the shelf at any sporting goods store and immediately use with limited prior knowledge. While highly accessible, their limited range may mean communicating only within your immediate vicinity.

GMRS—General Mobile Radio Service

The General Mobile Radio Service (GMRS) is another set of frequencies the FCC has designated for personal use. Unlike FRS channels, GMRS channels require a license, allowing for higher power levels and greater range. Highlights include (*General Mobile Radio Service [GMRS]*, n.d.):

- GMRS provides longer-range licensed two-way radio communication.
- It requires a simple FCC license (currently $35), which covers you and your immediate family for five years.
- Power output is limited to 50 watts for GMRS channels, with power limited to 0.5 watts for the channels in which GMRS and FRS overlap.
- Operates on 30 channels near 462 MHz and 467 MHz, with eight channels dedicated to repeaters exclusive to GMRS.
- The typical handheld range is under 5 miles, but repeaters can extend it further.

The GMRS license offers affordable, licensed, long-range communication for families and small businesses. The license gives you legal access to higher power levels and repeater infrastructure, allowing expanded coverage compared to unlicensed bands. For example, GMRS could enable clear communication across a large campground or construction site. It provides an ideal emergency channel for coordinating with family. The licensing also avoids interference issues common on crowded, unlicensed frequencies.

FRS vs GMRS – What's the Difference?

The FRS (Family Radio Service) and GMRS (General Mobile Radio Service) are radio services operating in the same 462-467MHz range of the UHF band. They share 22 simplex channels, meaning that a channel on an FRS radio corresponds to the same channel on a

GMRS radio across different brands. This compatibility allows FRS and GMRS users to communicate directly with each other.

However, there are key differences. FRS radios don't require a license and are restricted to handheld models with non-detachable antennas and a maximum of 2 watts of power on certain channels. In contrast, GMRS radios need a license. GMRS radios can be handheld, mobile, or base station units with detachable antennas and higher power limits – up to 5 watts on channels 1-7 and up to 50 watts on channels 15-22. GMRS also offers 8 additional repeater channels, which FRS doesn't have.

While FRS and GMRS share channels and frequencies, their distinct features, like radio types, wattage, antenna flexibility, and repeater capability, cater to different user needs. FRS is more suited for personal, casual use without the licensing hassle. In contrast, GMRS offers enhanced features and power for a more robust communication experience, albeit with a licensing requirement.

MURS—Multi-Use Radio Service

MURS, or the Multi-Use Radio Service, is a private, two-way, short-distance voice or data communications service for personal or business activities in the United States. MURS facilitates two-way communication without needing any license. Key details are (*GMRS*, n.d.):

- Operates on 5 VHF frequencies near 151–154 MHz.
- Limited to a maximum of 2 watts transmitter power output.
- It is intended for sporadic, short-distance exchanges.
- It has a typical range of 1–3 miles.
- MURS is used for personal and business communications by families, small businesses, and recreational activities. It's also used for remote control of devices and telemetry applications.

MURS provides an easy entry point into VHF radio operation for personal or business use without any license requirement. MURS allows doubled power levels, extended range, and business usage compared to FRS radios. The frequencies are less crowded than FRS as well. However, the range is still limited compared to licensed services like GMRS or commercial radios. MURS is a good option for simple, short-range communication at low cost and complexity. However, a licensed service may work better for heavy daily use or wide coverage.

Marine VHF Channels

These frequencies are expressly set aside for maritime communication, similar to dedicated boat lanes in a bustling harbor. These channels are used for various purposes, including communication between vessels, shore stations, and emergency distress calls.

If you're using your Baofeng radio in a marine environment, knowing how to access these Marine VHF channels can be a lifesaver. Whether calling for assistance in an emergency, receiving weather updates, or simply coordinating with other vessels, these channels are your lifeline in the open sea.

Aircraft Radio Service

The Aircraft Radio Service, essential for aviation safety, operates primarily in the VHF band (108-137 MHz) for communication and navigation. Managed by federal authorities, it facilitates crucial two-way communication between pilots and air traffic control, vital for flight operations, especially during takeoff, landing, and controlled airspace. This service includes emergency frequencies (121.5 MHz and 243.0 MHz). Operators require licensing, often included in pilot certifications, underscoring the service's role in coordinating global air traffic and maintaining high safety standards.

Amateur Radio Service

Amateur Radio, or "ham radio," is a hobby in which enthusiasts use various communication modes, such as voice, digital, Morse code, and slow-scan television, to connect worldwide. It's not just for casual chatting; ham radio operators also provide vital communication during emergencies when conventional networks fail. One must pass a licensing exam covering radio theory and regulations to participate. It's a diverse and engaging hobby, regulated by national authorities like the U.S. FCC to ensure responsible and safe use of the radio spectrum.

Here are some basics (*Amateur Radio Service*, 2016):

- It requires passing an exam to obtain an FCC license such as Technician, General, or Extra Class (more on these later). Each license level requires passing an examination, with the complexity and depth of the test material increasing with each higher level. The higher the class of license, the more operating privileges the holder has.
- It privileges larger frequency bands and power limits, dependent on license class.
- An estimated 730,000 licensees are in the US.

While licensing is a hurdle, ham radio offers extensive customization and long-range communication capabilities for electronics hobbyists. Compared to GMRS, ham radio allows greater technical experimentation and broader frequency access, but passing exams is required to unlock these privileges. GMRS provides a simpler entry point for affordable licensed long-range communication without customization.

In summary, choosing a radio service involves balancing licensing needs, intended use, power, and range requirements.

RADIO TECHNOLOGY FUNDAMENTALS

Now that we have covered radio frequency basics and services let's demystify some key radio technology terms to round out your knowledge foundation.

Radio Frequency (RF)

- Radio Frequency (RF) is vital in wireless communications. It encompasses electromagnetic waves for data transmission, with frequencies ranging from 20 kHz to 300 GHz. These waves enable technologies such as mobile phones, radios, and Wi-Fi.
- RF works by emitting waves that carry information through space, similar to ripples from a stone thrown in water. This principle allows voice, data, and video to be transmitted wirelessly over distances.
- RF is ubiquitous in modern life. It's essential for mobile communications, where voice signals are transformed into RF waves, connecting people without physical wires. For ham radio enthusiasts, like Baofeng radio users, understanding RF's properties is key for effective communication and troubleshooting.

In summary, RF is the cornerstone of wireless connectivity, crucial for the functioning of various modern technologies.

Frequency Modulation (FM)

FM radio works by changing or modulating the radio signal's frequency to encode information. It is like tuning a radio dial up and down to different stations. The frequency shifts up and down based on the loudness of the audio signal. FM encodes sound differently than AM radio.

Compared to AM, FM has some significant advantages:

- Reduced noise and interference since only one frequency carries the whole signal.
- Allows transmitting higher fidelity audio than AM modulation.
- Resists signal distortion from lightning, electrical motors, and power lines.

FM provides excellent audio clarity and resistance to external distortions for portable two-way voice communications, so services like FRS and GMRS use FM.

CTCSS and DCS

These systems help filter out unwanted conversations on a radio channel. They use unique low-frequency tones or codes sent along with the voice signal.

- The Continuous Tone-Coded Squelch System (CTCSS) transmits one of 50 super low, sub-audible tones. Sub-audible means below what humans can hear—under 20 Hz. You can't hear these tones but can feel them as vibrations! Each tone is unique (Vivian, 2021).
- Digital Code Squelch (DCS) uses 1 of 105 special digital codes sent with the voice signal (Vivian, 2021).
- Both radios must use the same CTCSS tone or DCS code to talk to each other. This blocks out other random conversations on that channel that do not match your tone or code. It's like a password or filter to keep communications more private.

These extra tones and codes help prevent unwanted chatter from cutting in. You only hear radios using the right tone or code to match yours. It clears up the channel, so you get the calls that matter to you.

Antennas and Transmission

Antennas take the incoming radio waves and convert them into electrical signals, which are fed into the receiver (Ho, 2023). They also take transmission signals from the transmitter and convert them into radio waves. Key aspects are:

- Efficient power transfer between transmitter/antenna and antenna/receiver is critical for maximum range.
- Antenna gain in the desired direction results in higher signal strength.
- Matching the antenna design parameters like frequency, impedance, and radiation pattern to the radio frequency is vital for optimal performance—for example, a VHF antenna would be poorly matched to a UHF radio.
- Environmental factors like terrain and structures impact the propagation of transmitted signals.

So, upgrading to a better antenna significantly improves performance. We will cover optimal antennas for Baofeng radios later in the guide.

Understanding how antennas radiate signals and interact with the environment further demystifies radio wave transmission and reception.

We will build on these radio basics extensively in later chapters. However, you are now equipped with a foundational understanding of radio. Next, look at what features are important when buying a Baofeng.

DECIPHERING THE FEATURES OF RADIOS

With the wide range of Baofeng two-way radio models available, it can be overwhelming trying to understand all the technical specifications and features. However, understanding the key capabilities and differences between models is essential for making the right purchase decision tailored to your needs. We will break down the most critical features to evaluate across popular models so you can zero in on your ideal device.

Range

One of the most vital features to consider is the expected transmission range (Encyclopedia Britannica, Inc., 2022). After all, the range determines the maximum distance your messages can cover and receive replies from other users. Baofeng radios support frequency bands like VHF and UHF, which have innate differences in propagation characteristics that impact range (Chen, 2021):

- VHF signals can travel further distances in outdoor open environments thanks to line-of-sight propagation. However, the range is more limited indoors due to attenuation (the reduction in strength of a radio signal as it travels through things such as walls and objects).
- UHF signals, while having a reduced range in open areas compared to VHF, can better penetrate through foliage, buildings, and other obstructions. This makes UHF well-suited for dense urban settings.

Within VHF and UHF bands, several factors like transmission power output and the antenna used significantly influence the range (Encyclopedia Britannica, Inc., 2022). Most handheld Baofeng models output 5W of power, but high-powered units like the BF-F8HP can transmit 8W for extended reach. Similarly, upgrading to an after-market high-gain antenna can dramatically increase the range compared to the standard rubber duck antenna.

However, when comparing the different Baofeng models for the expected range, the tested max range under optimal conditions provides a helpful indicator.

Entry-level models like the UV-5R are rated for up to 5 miles under ideal circumstances using the stock antenna (*Baofeng UV-82HP vs UV-5R*, 2022).

Upgraded models like the BF-F8HP can achieve over 15 miles of range thanks to their higher transmit power and better internal components (*Best Baofeng Radios*, 2022).

Due to power restrictions, commercial variants like the BF-888S have a more limited range of just 1–2 miles (TalkieMan, 2019b).

Consider your typical usage—a basic entry-level model may suffice if you need short-range communications within a few blocks in urban areas. However, the extended reach of high-powered radios like the BF-F8HP will prove invaluable for remote locations with challenging terrain.

Battery Life

Baofeng handhelds rely on rechargeable Li-ion battery packs for power. Enough battery capacity to last through the day is essential when using your radio for outdoor activities or emergencies. Entry-level Baofeng models often have limited battery life, especially on high-power transmissions. Upgraded models feature improved batteries offering enhanced runtime between charges (Knoji Staff, n.d.; Superior Support, 2022):

- The UV-5R ships with a standard 1,800 mAh battery, providing only 8–10 hours of battery life. Heavy use will drain it within half a day.
- In comparison, the UV-82HP has a beefy 2,800 mAh battery, providing 15-18 hours of operation and allowing you to use the radio from sunrise to sunset without recharging.
- Some premium models, like the UV-9R Plus, boast extended 2,500 mAh batteries despite being compact. These models strike a balance between battery life and portable size.

If you need your radio for all-day outdoor excursions or camping trips or as an emergency preparedness tool, prioritize those Baofeng models from the factory that offer the highest capacity battery packs. Most batteries can be replaced with third-party options. However, it's convenient when the stock battery provides sufficient operating time before needing spares.

Durability and Water Resistance

Baofeng radios are inherently durable, given their compact polycarbonate casing and sturdy controls. But you can expect specific models to withstand the elements better when used outdoors:

- Entry-level models like the UV-5R lack any water resistance. Exposure to heavy rain or submerging them can permanently damage the internals (Rock Networks, n.d.-a).
- Upgraded variants like the UV-9R Plus offer an IP67 rating, making them fully dustproof and waterproof up to 1-meter depth for 30 minutes. Their rubber gaskets provide a sealed design (*Best Baofeng Radios*, 2022).
- The UV-82HP features a more robust chassis than the UV-5R and uses metal alloy in high-wear parts. However, it lacks a sealed design, so water should be avoided (Superior Support, 2022).

If your radio will likely get wet when used in inclement weather or take the occasional splash during water sports, opting for one of the IP67-rated waterproof models is wise. Similarly, choose a dustproof-rated option if it will be exposed to dirty or dusty environments on your adventures. However, the added weather resistance comes at a higher price. If you plan to use your radio only in fair weather, an entry-level model like the UV-5R will likely suffice since it will be affordable to replace if damaged.

Audio Quality and Volume

Being able to hear messages clearly is vital to effective communication. Some environments, like noisy construction sites, require loud and clear audio. Evaluating the speaker quality and volume levels across Baofeng models is essential if audio clarity is paramount (Rock Networks, n.d.-a):

- The speaker on basic models like the UV-5R maxes at around 700 milliwatts (mW), a relatively low power level. In noisy surroundings, received voices can be nearly unintelligible, even at full volume.
- Premium models, like the UV-82HP, boast 2-watt (W) front-facing speakers. Combined with the dual push-to-talk (PTT) feature, they excel in noisy workplace conditions with loud incoming audio.
- For discrete communications, personal receiver earpieces allow hearing transmissions clearly without blaring the message aloud. Most Baofeng radios support external microphones and earpieces.

Speaker capabilities may be less critical if you only plan to use your radio occasionally in quiet indoor settings. However, opt for a loud, clear audio model for outdoor group hikes or crowded public events to avoid miscommunications.

Ease of Use

All Baofeng radios feature a standard control interface with a PTT button (Push to Talk), volume knob, keypad, and LCD screen. However, some models make the experience more ergonomic and intuitive (Rock Networks, n.d.):

- Models like the UV-5R have a smaller monochrome display and buttons that can feel cramped.

- Newer designs like the UV-9R Plus feature a larger 1.5" color display, oversized PTT, and a more ergonomic grip for one-handed use.
- The UV-5R Plus simplifies the interface by adopting an Android smartphone-style menu UI (user experience) rather than the traditional icon-based menu navigation.

If you are less technically inclined or want an intuitive radio that even beginners can operate, evaluate the ease of use. Handling the device and comparing the button sizes, display visibility, menu navigation, and grip ergonomics in person can be worthwhile. While all Baofeng radios share a common architecture, subtle refinements across models impact usability.

Programming Capability

One of the top advantages of Baofeng radios is their programmability using a computer and programming cable. This allows configuring them for your desired frequencies and channels beyond the default settings. However, specific models are more accessible to program than others:

- The UV-5R is notoriously tedious to program manually, requiring combing through confusing menus. Software like CHIRP simplifies the process, but some users still experience hiccups (Centers, 2022).
- Newer options like the UV-9R Plus can be programmed directly on the device. Adding channels with the intuitive interface is more manageable, reducing reliance on a computer.
- Some premium models even include built-in touchscreens allowing programming from the radio without cables.

Those who plan to minimize manual programming and use default frequencies may find even basic Baofeng models sufficient. But if you intend to program memory banks spanning multiple channels, opt for a model that streamlines and simplifies the process.

ACCESSORIES

One benefit of the immense popularity of specific Baofeng models is the wide range of compatible accessories (Encyclopedia Britannica, Inc., 2022). You can enhance your radio with add-ons like:

- Aftermarket antennas: Upgrade your whip or duck antenna for increased range with options like the Nagoya NA-771.
- Headsets: Discreetly receive transmissions using earpieces. Specific models have built-in headset jacks.
- Microphones: Use extended microphones for long-range or outdoor use, keeping the radio itself protected.
- Batteries: Extended batteries help keep the radio operational for multi-day trips.
- Belt clips: Keep the device securely attached and instantly accessible.

Well-known models like the UV-5R, BF-F8HP, and UV-82HP have a thriving accessory ecosystem. Some key points when evaluating accessories:

Antennas

- Ensure any aftermarket antennas are designed for your radio's frequency bands with appropriate connectors. Also, match the antenna impedance to the radio's rated impedance (the resistance of the circuit - impedance is crucial in radio systems for matching the transmitting antenna and the receiver to ensure efficient power transfer and minimize signal reflection and loss).
- Higher gain (directional antennas) antennas boost range but require compatible power levels supported by the radio.

- Select antennas suited for your usage, like flexible rubber models for outdoors or small magnetic mounts for vehicular use.

Cases and Clips

- Hard cases offer the most protection, while silicone and TPU (Thermoplastic Polyurethane) cases are more grippy. Soft pouch cases make radio storage and transport convenient.
- Belt clips must match the radio's body in tightness, width, and mounting style to stay securely attached.
- Commonly available belt clips usually work for the UV-5R and UV-82. However, the BF-F8HP requires specially sized clips that fit its thicker body.

For a safe, customized setup, ensure all add-ons and replacements meet the technical specifications of your particular radio model.

UNDERSTAND THE DIFFERENT BAOFENG MODELS

Having covered the key factors, like frequency, power, and programming, let us explore some specific Baofeng models with those features to help clarify choosing the right radio. With so many Baofeng models available, it can get overwhelming trying to determine which one is right for you. Each model has its unique blend of features, performance factors, and price points. Evaluating your specific needs and usage scenarios will help narrow down the ideal option. We will provide an in-depth feature comparison across some of the most popular Baofeng radios to simplify your purchase decision.

Baofeng UV-5R

The UV-5R is the classic Baofeng model, the gold standard of affordability and functionality. It operates on VHF 136-174MHz and UHF

400-520MHz frequencies with a power output of 5 watts. The UV-5R is compact and lightweight (about the size of a deck of cards), making it a great companion for outdoor adventures. It includes features such as dual-band, built-in flashlight, and an FM radio, which come in handy when you least expect it.

The UV-5R is feature-packed, given its low cost, making it a versatile starter radio. But the maximum 4W power output and short battery life limit its utility for long-range or all-day use. Still, it remains a popular "first Baofeng" choice.

Baofeng UV-82HP

The UV-82HP model builds on the UV-5R, improving some drawbacks for a modest price increase. The UV-82HP is a model of choice for many radio enthusiasts. It operates on the same frequencies as the UV-5R and BF-F8HP but has a power output of 7 watts. The higher capacity 2800 mAh battery provides 18+ hours of battery life, over double the UV-5R (Superior Support, 2022). What sets the UV-82HP apart is its ergonomic design. The radio's button layout is designed for ease of use, and the built-in speaker delivers crisp and loud audio.

So, suppose you need a rugged radio with extended runtime suitable for outdoor conditions or workplace use. In that case, the UV-82HP is worth the slight cost increase over the UV-5R. Just lack of weatherproofing is a limitation.

Baofeng BF-F8HP

Where the UV-82HP improves battery life over the UV-5R, the BF-F8HP boosts power output. The BF-F8HP is an upgrade over the UV-5R, boasting a power output of up to 8 watts. The increased power means better range and signal clarity, especially in challenging environments. The BF-F8HP also operates on VHF 136-174MHz and UHF 400-520MHz frequencies. It shares many features with the

UV-5R, but its standout quality is its improved battery life, thanks to the 2100mAh battery. It also has an IP55 rating (see the glossary for an expanded definition of IP ratings, but it measures protection from solid objects and liquids), which provides some dust and splash resistance (*Baofeng - Baofeng Troubleshooting*, 2022)

On the other hand, the BF-F8HP does not have an illuminated keypad or shortcut keys. Moreover, it is about 25% heavier than other handheld models (Best Baofeng Radios, 2022). If you prioritize extended-range capability, the BF-F8HP is a top choice despite its larger form factor.

Baofeng UV-9R Plus

The UV-9R Plus is Baofeng's attempt at a more premium model, boasting an IP67 waterproof and dustproof rating. It operates on VHF 136-174MHz and UHF 400-520MHz frequencies with a power output of 8 watts and a longer-running 2500 mAh battery. The UV-9R Plus is a fantastic choice for outdoor adventurers, as it can withstand harsh weather conditions. This model also features a dual push-to-talk button, allowing you to switch between two frequencies easily.

Due to the upgrades, the UV-9R Plus is more expensive than the UV-5R. However, if you need a rugged radio with excellent battery life, the UV-9R Plus checks all the boxes.

Baofeng BF-888S

The BF-888S is Baofeng's entry in the compact FRS/GMRS walkie-talkie category (Rock Networks, n.d.; TalkieMan, 2019). Attributes include a smaller, compact design ideal for a backpack or pocket. It supports 2W output power and a basic range of 2–5 miles. The 16 channels in FRS/GMRS frequencies of 462 and 467 MHz are great for something that needs to work right out of the box with no license required.

This micro-sized option is designed for casual short-range communication. It provides an affordable, lightweight walkie-talkie option.

MAKE YOUR SELECTION

Like choosing between different types of vehicles, choosing the suitable Baofeng model depends on your specific needs and circumstances. For example, if you're going for a hike on a clear day, the UV-5R might be enough. But if you're planning a week-long camping trip in unpredictable weather, the weather-proof UV-9R Plus could be your best bet.

Remember, the perfect radio doesn't exist, but the ideal radio for you does. Understanding the features and capabilities of each Baofeng model can help you make an informed decision.

Key Factors in Selecting a Baofeng Radio:

- Range: Models like the BF-F8HP offer extended reach due to higher power output.
- Battery Life: Consider models like the UV-82HP and UV-9R Plus for longer battery endurance.
- Durability: The UV-9R Plus stands out with its IP-rated water and dust resistance.
- Audio Quality: The UV-82HP is notable for its superior speaker output.
- Ease of Programming: While all models are programmable, some may offer a more user-friendly experience.
- Portability: The BF-888S is compact and lightweight, perfect for on-the-go use.

With this knowledge, you can confidently select a Baofeng model that delivers the right mix of functionality and value tailored to your communication needs.

Making the Right Purchase: When buying your chosen Baofeng radio, you have several options, each with advantages and considerations.

Buying Online: Online platforms like Amazon offer a wide selection of new and used Baofeng radios.

Baofeng Specialty Retailers:

- Direct purchases from sites like BaoFengTech.com guarantee genuine products and full warranty support.
- Specialty retailers like Buy Two Way Radios and STS Telecom offer expert advice and a focused selection of Baofeng products.

General Electronics Stores:

- Stores like Best Buy or Walmart may carry Baofeng radios, offering the advantage of in-person purchases.
- The drawback is that staff may lack in-depth knowledge of Baofeng radios, and selection might be limited.

Local Ham Radio Outlets:

- Local stores often have knowledgeable staff with practical experience in ham radio.
- Finding a well-stocked outlet can be challenging, especially in smaller towns.

Purchasing Considerations:

- Check for warranty coverage and ensure it's from an authorized dealer.
- Understand the return policy.
- Confirm if the seller offers post-purchase technical support.

Avoiding Counterfeit Baofeng Radios:

- Buy only from authorized dealers listed on official Baofeng websites.
- Confirm the FCC ID on the radio matches the FCC's database. Counterfeits usually lack proper FCC certification.
- Check for FCC certification on the radio and verify it on the FCC ID search website (https://www.fcc.gov/oet/ea/fccid).
- FCC Equipment Authorization

  - Just as a new car must meet specific safety and emissions standards before it can hit the road, radio transmitting equipment like your Baofeng radio must meet specific technical standards the FCC sets before it can be marketed and sold in the United States. This is known as FCC equipment authorization.
  - As a user, you can check if your Baofeng radio is FCC-authorized by looking for an FCC certification label on the device or by searching the FCC's Equipment Authorization Database.

- FCC ID and Labeling Requirements

  - Imagine walking into a library and finding all the books without titles or authors. It would be nearly impossible to find what you're looking for. Similarly, the FCC requires all authorized radio devices to have an FCC ID. This unique identifier helps the FCC keep track of all devices in the market and allows consumers and the FCC to retrieve the device's certification information.
  - The FCC ID is typically found on a label affixed to the device. This label must be permanently attached

and readily visible. It includes the FCC logo, the FCC ID, and a statement confirming the device's compliance with the FCC's technical regulations.

Retaining Important Documentation:

- Keep all provided documentation, including the user manual, warranty card, packing list, and FCC grant.
- These documents are essential for setup, warranty claims, and verifying FCC compliance.

Understanding these aspects of purchasing a Baofeng radio will help you navigate the market effectively. Whether you choose a specialty retailer for expert advice or an online marketplace for convenience and selection, ensuring the authenticity and suitability of your radio is paramount. With careful consideration, you can acquire a Baofeng radio that perfectly fits your communication needs and skill level.

FINAL CONSIDERATIONS: ADD ONS

Choosing the best Baofeng radio model is just the beginning of your journey. To truly maximize the potential of your new radio, it's important to consider investing in essential accessories. These additions can significantly enhance the functionality, durability, and overall experience of using your Baofeng radio.

Essential Baofeng Radio Accessories:

1. Spare Batteries:

- Crucial for extended use, especially in areas without charging facilities.
- Vital in emergencies where power sources may be scarce.
- Cold weather can drain batteries faster, so having spares is beneficial.

2. Battery Chargers:

- Multiple chargers ensure you can recharge batteries wherever you are – at home, in your vehicle, or in a bug-out bag.
- Owning more than one charger provides backup in case one fails.

3. Carry Cases:

- Protects your radio from physical damage.
- Waterproof cases guard against rain and moisture.
- Features like belt clips and storage pouches enhance convenience and organization.

4. Microphones and Headsets:

- Ideal for hands-free operation and discreet communication.
- Improves audio clarity, especially in noisy environments.
- External PTT buttons facilitate easier transmission.

5. Upgraded Antennas:

- Enhances reception and transmission range.
- Select an antenna compatible with your radio's frequency range and suitable gain rating.
- Durability and size are important considerations based on your usage.

6. End Fed Half Wave Antennas:

- For long-range communication in emergencies.
- Portable and quick to deploy, they can significantly extend your radio's reach.
- Dual-band models are versatile for various frequency needs.

7. Programming Cable:

- Facilitates easy programming of channels via software like CHIRP.
- Essential for backing up settings and updating firmware.
- Opt for genuine Baofeng cables for the best compatibility.

8. AA/AAA Battery Pack:

- Adds flexibility when standard batteries are more readily available.
- Useful in prolonged scenarios where recharging Li-ion batteries is not an option.

9. Go Bag:

- A tactical bag for storing and transporting your Baofeng and accessories.
- Features like MOLLE webbing and internal organization cater to various gear.
- Ensures readiness and quick deployment in any situation.

Once you've acquired your Baofeng radio and essential accessories, the next step is unboxing and setting it up. The upcoming chapter will guide you through this process in a beginner-friendly manner, covering:

- Initial unboxing and inspection of the radio and its components.
- Basic setup procedures, including battery installation and charging.
- Programming your first channels using the keypad.
- Testing key functions to ensure your radio is ready for use.

Even if you're new to radio communications, this next chapter will provide clear, step-by-step instructions to get your Baofeng radio up and running. Get ready to embark on an exciting journey of discovery and communication with your new Baofeng radio!

2

BREAKING INTO BAOFENG - UNBOXING AND BASIC OPERATIONS

The arrival of new gear, particularly something as crucial as a communication tool, is a moment charged with anticipation and excitement. Unboxing a Baofeng radio isn't just about revealing a new gadget; it's about unveiling a key component of your preparedness arsenal. As you tear through the packaging, it's not just the thrill of a new possession but the realization that you're a step closer to being better prepared for emergencies.

This Baofeng radio represents more than a communication device; it's your lifeline to the outside world in times of crisis, your tool for connecting with members during drills or actual emergencies. Its significance is profound for someone dedicated to preparedness. Every dial, button, and antenna is a component in your readiness strategy, ensuring that you have the means to stay informed and connected whether you're facing a natural disaster, a power outage, or any unforeseen event.

So, as your heart races with excitement over this new acquisition, know that it's more than just an addition to your tech collection; it's a vital step in fortifying your preparedness plan.

THE FIRST ENCOUNTER: UNPACKING YOUR BAOFENG RADIO

The excitement of unboxing a new gadget! As you unwrap your Baofeng radio packaging, here's what you'll typically find inside (Superior Support, 2022):

The radio device itself: Most Baofeng radios have a compact, rectangular shape that fits nicely in your palm. Popular models like the UV-5R are about 5.5 inches tall and 2 inches wide (again, about the size of a deck of cards).

- **Keypad:** The keypad is your control center, your command bridge. It's where you input frequencies, navigate menus, and access the various features of your radio. Each key on the keypad has a specific function beyond just number input. For instance, the [#] key will lock the keypad if pressed and held for 2 seconds (press and hold for 2 seconds to unlock it). The more comfortable you become with the keypad, the smoother your radio operation.
- **Display Screen:** The display screen is your window into the workings of your Baofeng radio. It's where you can see the current frequency or channel, monitor transmission status, and access menus. It also displays battery status and other settings like power and squelch levels. The display screen is backlit, making it easy to read even in low-light conditions.
- **Speaker/Microphone**: Are your radio's voice and ears. The speaker broadcasts the incoming signals, turning them into audible sounds. Conversely, the microphone captures your voice, converting it into a signal that can be transmitted. These components are usually combined into a single unit in handheld radios like Baofeng radios.

Battery: Your radio's battery is its heart, pumping the power needed to function. Baofeng radios typically come with a rechargeable

Lithium-ion battery, known for their long lifespan and high energy density. This means you can operate your radio for a considerable time before recharging. To install, match up the alignment guides on the battery to the radio and slide the battery upwards until it clicks securely. To remove it, turn your radio unit to the back. You'll notice a small latch on the back at the top. Push the latch down and slide the battery down and out.

Battery charger: While Li-ion batteries are great, they eventually drain. The battery charger restores the health of your battery heart! The battery charger typically comes with a USB cord that can be inserted into either a wall socket or a car charger.

- Batteries can be charged alone in the battery charger. You do not need the battery attached to the radio to charge it. This becomes important when you use your radio and have a backup battery in the charger, ready for future use.

Antenna: Think of the antenna as the periscope of a submarine. The component reaches into the vast ocean of frequencies, catching signals transmitted in its range. A standard Baofeng radio comes with a rubber-duck antenna, which is versatile, portable, and suitable for most situations. It's crucial to remember that the antenna is the life-line of your radio, the primary tool for receiving and transmitting signals. Treat it with care, and it will serve you well.

- Attach the supplied rubber antenna to the SMA (SubMiniature version A) female connector protruding from the radio's top. Screw it on gently but firmly until snug (Rick, 2016).

User Manual: This short and sweet document provides instructions for setup and operation. Keep it handy as you get started. While it is limited in scope and explanations, it is still helpful and convenient.

Belt Clip: A plastic clip lets you attach the radio to your belt or strap. If you regularly carry your radio and like the feel of it on your belt, I recommend upgrading to a sturdier carrier.

Wrist Strap: A loop that attaches to the radio and helps prevent dropping while carrying the radio. There's not much else to say about this!

Programming Cable (with some models): Connects your radio to a PC for programming. Imagine trying to input a hundred different contact numbers into your smartphone manually. Tedious, isn't it? That's where the programming cable comes in. This accessory connects your Baofeng radio to your computer, allowing you to use software to input frequencies, name channels, and customize other settings. Say goodbye to the monotony of manual input and hello to the efficiency of the programming cable.

Earpiece with Microphone (with some models): Allows hands-free talking. The external microphone is an accessory that you may find helpful for hands-free communication. Think about those times when you're multitasking, maybe cooking dinner while coordinating a neighborhood watch. An external microphone allows you to relay messages without holding the radio. Some models even come with a built-in speaker, enhancing the audio clarity.

———

Once you've inventoried the contents, installed the battery, and connected the antenna, your radio is ready for that exciting first power-on! So go ahead and explore your Baofeng radio. Hold it in your hands and feel its weight and texture. Familiarize yourself with its parts and features because this is not just a radio. It's your companion on a journey of discovery and into the world of radio communication. And as with any journey, the first step is always the most exciting. So, let's take that step together. Your Baofeng radio is ready. Are you?

POWERING ON: INITIAL SETUP OF YOUR BAOFENG RADIO

After attaching the battery and antenna, the moment you've been waiting for has arrived —you can power on your radio for the first time (*How to Use Baofeng Radios*, 2022).

Turn the radio over and locate the volume/power knob on the top. Turning this knob clockwise powers the radio on. You'll hear a beep and see letters or graphics displayed on the screen, indicating it is powered on. If the voice prompt is turned on, you will hear if you are in frequency or channel mode. The LED indicator will also light up, confirming the radio is now on.

On some models, you can customize the initial setup's radio name, owner name, power-on message, and menu language directly from the keypad. However, other models require you to connect to a computer using the programming cable.

GETTING FAMILIAR: UNDERSTANDING THE BASIC OPERATIONS

Now that your Baofeng radio is powered on and initialized, it's time to understand the basic controls and functions. Getting familiar with buttons and knobs will help you operate the device confidently (*How to Use Baofeng Radios*, 2022).

On the top, you'll find the volume/power knob, which turns the radio on and off and controls its audio volume. Turn clockwise to increase loudness.

The Push-to-Talk (PTT) button is on the left side and has an indent for easy identification by touch. Pressing and holding this button transmits your voice at the set frequency.

- Also, on the left side is a bright orange CALL button. Go ahead and push that button – but push it and let go. Please don't hold it unless you want to set off the alarm (more on the alarm later)! You turn on the FM radio by just pushing and releasing the button! Then, tune into your favorite FM radio music station using the number buttons. Push and release the button to turn off the FM radio.

- The Final button on the left is MONO. For now, push that button and see the flashlight come on. Press again for a strobe light, and press again for off. The other function of this button will be covered later in the book.

The numeric keys for inputting frequencies and text are on the front of your radio. The keys are also used for selecting menu options and programming the radio.

- The menu button activates the radio's menu system. Use it to access configuration options and settings. The "MENU" key also acts as the "enter" function. When you want to select a menu option, click the menu button. The arrow keys help navigate between menus and settings. The basic UV-5R has 41 menu sections (menu sections 0-40)!
- Also seen on the front is the transmit/receive LED indicator. This is activated when transmitting or receiving signals. It blinks green when receiving and turns solid red when transmitting. Within the menu, these are slightly customizable, with the ability to select between a few color options.
- The bright orange button is the VFO/MR. This button switches between 'frequency mode' and 'channel mode.' VFO (Variable Frequency Oscillator), or frequency mode, allows you to enter a frequency manually. MR (memory recall), or memory mode, will enable you to access your saved channels easily.
- Also standing out on the front in a nice blue color is A/B. This button switches between the two frequencies displayed on the screen. Go ahead and press A/B, and watch as the arrow

moves between the top and bottom frequencies. We'll cover their usage in more detail later. For now, familiarize yourself with button placement and functionality.

LET'S TALK: SETTING UP YOUR FIRST COMMUNICATION

Once you have a basic grasp of the controls, it's time to set up your first actual two-way communication using your Baofeng radio. This helps confirm everything is working correctly and builds confidence in using your radio (*How to Use Baofeng Radios*, 2022).

First, ensure the battery is fully charged—the icon on the display indicates the battery level. Then, turn the radio on!

Set your radio to a frequency like FRS channel 15 (462.550 MHz). Use the numeric keypad to input the 6-digit frequency. Have a second person ready with their own FRS or GMRS radio set to the same channel or frequency.

Press and hold the PTT button on your Baofeng radio, speak into the mic to transmit your voice message, and then release PTT when done. The other radio should receive your audio through its speaker. Have the other person press their PTT and respond to confirm two-way communication.

Start by testing at close range, then slowly increase the distance between the two radios while communicating to determine the maximum range. Try moving between, in, and around different structures or buildings. Keep your Baofeng's antenna vertical for best results. This test confirms everything is working and teaches you about practical radio communication.

Now, let's say you don't have someone or another radio. Then, hold tight, and we will connect you with the NOAA Weather station for your area.

Find your local NOAA weather radio frequency:

- Go to www.weather.gov/nwr and click the "Find My Station" button.
- Enter your zip code into the 'NWR Station Search' section and press 'Enter.'
- Note the frequency in the yellow box. In my area, the frequency is 162.550 MHz.

On your radio, first, ensure you have selected the correct band. Click the BAND button until you are on VHF (136–174 MHz range). To help me, I think of 'V' in VHF as standing for very long distances (excellent outdoors over open terrain). Since your weather station is probably a long way off, you want to be on VHF.

- Type in the frequency of your weather station.
- You may have to wait a minute to hear the transmission.
- You may also hear static or breaks in the communication. If that is the situation, hold MONO on the left side of the radio press. We will explain this function later, but it will help you hear the transmission for now.

You did it! You are now on the radio. You have taken your first step to becoming a radio operator and setting yourself up for future preparedness. Congratulations!

3 or 4 Decimal Places

When programming frequencies into your Baofeng, you might encounter a situation where the frequency you need to input has more decimal places than your radio seems to support. Most Baofeng radios, including the UV-5R, typically display and operate with frequencies up to three decimal places (e.g., 162.555 MHz).

If you need to access a frequency with four decimal places (like 162.5575 MHz), here's what you can do:

- Round to the Nearest Frequency: In your case, you could round 162.5575 MHz to 162.558 MHz or down to 162.557 MHz. The Baofeng UV-5R should accept either of these frequencies. The slight difference is often within the acceptable tuning error margin for most amateur radio and scanning purposes.
- Consider the Bandwidth: The bandwidth of your transmission (or the receiver's filter bandwidth) may be wide enough to accommodate slight frequency discrepancies. For instance, if you're using a typical FM mode with a bandwidth of 5 kHz or more, being off by a few hertz will generally not be significant.
- Check Real-World Performance: Test the radio in real-world conditions after rounding the frequency. If you're trying to access a specific repeater or listen to a particular signal, see if the radio successfully receives or transmits as expected. Often, you'll find that the minor discrepancy doesn't impact functionality.
- Use Software for Programming: Programming software like CHIRP might allow you to enter the frequency more precisely. However, the radio's actual tuning accuracy and display will still be limited to hardware capabilities.
- Hardware Limitations: Remember that the Baofeng UV-5R and similar models are budget-friendly radios. Their frequency resolution and accuracy might not be as high as expensive amateur radio equipment.

For most amateur radio applications, especially on VHF and UHF bands, rounding to the nearest available frequency setting on your Baofeng should be sufficient. However, if you frequently need to operate with greater precision, you might consider investing in a radio with finer tuning capabilities.

RADIO ETIQUETTE

Now that you are "on the radio," let's review radio etiquette. This section will guide you through some of the most important ones.

Stay Legal

First and foremost, know what frequencies you are on and which ones you are legally allowed to transmit. If you are unsure, for now, just listen.

Listening First

Before you hit that transmit button, it's essential to listen first. Check if the frequency or channel is currently in use. Interrupting ongoing communications is considered rude in the radio community. Take the time to understand the flow of the conversation, and when you're sure the channel is clear, proceed with your transmission.

Call Sign Usage

Imagine attending a formal event where everyone is addressed by their titles and names. It lends a sense of identity and respect to the conversations. In the realm of radio communication, call signs serve a similar purpose.

A call sign is a unique identifier essential in radio communication, assigned to stations and operators for organized and legal frequency use. In amateur radio, the call sign, issued by authorities like the FCC in the U.S., comprises a region-indicating prefix, a local area numeral, and a unique suffix.

Operators must use their call signs at the start and end of transmissions, ensuring clear identification and responsible frequency use. Managed internationally by the ITU, these call signs facilitate the global identification of transmission origins. Some operators personalize their call signs, adding elements like initials. As a vital component of radio communication, call signs ensure order and accountability across radio networks.

Remember, your call sign is your identity on the airwaves. Use it with pride and respect.

Brevity and Clarity

Keep your communications short and to the point when speaking over the radio. Remember that the airwaves are shared, and lengthy transmissions can tie up a frequency. Speak clearly and slowly enough to be understood, but try not to hold the airwaves longer than necessary.

Using Standard Phrases and Protocols

Radio communication has its language and protocols. Using standard phrases and codes can make your transmissions easier to understand.

Staying Polite and Respectful

Remember to always be polite and respectful in your communications. Avoid using offensive language or engaging in arguments over the air.

RADIO LANGUAGE: Q-CODES AND PHONETIC ALPHABET

Use of Q-Codes

Imagine being in a foreign country, unable to speak the local language, relying solely on a handful of universally understood signs to navigate your way. In the world of radio communication, Q-codes serve a similar purpose. They're a set of standardized abbreviations used to convey specific messages, transcending language barriers and ensuring clear communication.

Your Baofeng radio allows you to communicate using these Q-codes. For example, if you're in an emergency and need assistance, you can use the Q-code 'QRR' to signal that you're in distress and require immediate help.

Phonetic Alphabet

Think back to your school days when you first learned the alphabet. Each letter had a name and a corresponding sound that helped you remember it. In radio communication, the phonetic alphabet functions in a similar way. It's a set of words used to represent each letter of the alphabet, ensuring clear and accurate communication over the airwaves.

The phonetic alphabet is particularly useful when communicating complex information, such as coordinates or license details, over your Baofeng radio. For example, instead of saying 'B,' which might be misheard as 'D' or 'V,' you would say 'Bravo.' This ensures that the information you're transmitting is received accurately, even in challenging conditions.

A printable version of Q-codes and the Phonetic Alphabet can be found here:

Scan the QR code or click the link to receive 'The Baofeng Radio Revolution Command Packet,' which includes a formatted and printable copy of Q-Codes and the Phonetic Alphabet. Print and keep close to your radio.

www.MorseCodePublishing.com/Command

COMMON ISSUES: NAVIGATING THE FIRST ROADBLOCKS

As with any new technology, there is a learning curve. You may encounter minor issues initially when using your Baofeng radio. Here are some common problems beginners face, along with potential solutions (*Baofeng - Baofeng Troubleshooting*, 2022; Rick 2016).

- Difficulty powering on: First, ensure the battery pack is charged and correctly inserted. Check that the power/volume knob is turned clockwise to switch it on. If it still doesn't power up, contact the dealer.
- No sound: Check if the volume is turned up adequately by turning the volume knob. Inspect the speaker to ensure it is clean and not blocked or damaged.
- Battery drains quickly: Turn down the brightness if it is too high. Disable features like VOX if not required to save battery life. Check the battery icon - if it shows just 1 bar, charge the battery.

These basic troubleshooting tips will help you resolve common initial hiccups. Don't be discouraged by minor issues. Regular usage will make you increasingly adept at using your Baofeng radio.

LEARNING CURVE: BECOMING COMFORTABLE WITH
YOUR BAOFENG RADIO

Developing familiarity and comfort with your new Baofeng radio
takes regular usage. Here are some tips to quickly become proficient:

- Practice operating all the buttons and looking through the
  menu until the layout and functioning becomes second
  nature. Muscle memory helps build reflexive skills.
- Initially, focus on key features like changing frequencies,
  transmission, and reception. Don't get overwhelmed trying
  advanced options before mastering the basics.
- Another quick and easy thing to start with is to scan
  frequencies. Here is how:

  - Go to Frequency Mode – click the VFO/MR
    button.
  - Click BAND and select either VHF or UHF. Try
    them both, but start with VHF.
  - On the UV-5R, click the * button with blue writing
    on it 'SCAN.' Press and hold until you hear
    'Scanning Begin,' or if your voice prompts are off,
    you see the frequencies start changing.
    Then release.
  - Your radio will scan through all available
    frequencies.
  - Listen and see what you pick up. Then, write down
    that frequency for future use.
  - Don't worry if you don't hear much. It can be hit or
    miss at times. But if you are scanning VHF, you
    should hear your NOAA weather station again.

In this chapter, we walked through unboxing your new Baofeng radio,
powering it on for the first time, understanding the essential functions, setting up your first communication, and getting comfortable

with operating it. Don't worry if you don't feel like you know every-thing yet—the key is just getting hands-on experience and becoming familiar with the basics. Now, you are ready to dive deeper.

———

The next chapter will help cut through the noise by explaining common radio terms and jargon you'll encounter with your Baofeng. Understanding these key terms and concepts will make mastering your radio much more manageable. You've completed the first step of unboxing and basic setup—get ready to boost your radio knowledge even further in the next chapter! Let's get started learning essential radio lingo.

3

DECODING BAOFENG - A BEGINNER'S GUIDE TO RADIO TERMINOLOGY

Mastering the language of your tools is as crucial as owning them. Baofeng radios, a staple in any prepper's kit, come with unique technical jargon and terminology. Understanding these terms is not just about technical knowledge; it's about ensuring effective communication during critical situations. Whether coordinating with a group during a drill, communicating during a community event, or navigating through an emergency, the clarity and precision afforded by proper terminology can make all the difference.

This chapter is dedicated to demystifying the language of Baofeng radios. It's designed to help you confidently operate your radio and communicate with precision and authority. By grasping these key terms, you're enhancing your ability to use your Baofeng radio as an efficient tool in your preparedness arsenal, ensuring that your communication is clear, concise, and effective when needed.

THE BASICS: THE IMPORTANCE OF PROPER RADIO TERMS AND LANGUAGE

You frequently encounter technical terms and acronyms specific to two-way radios, whether reading the user manual, browsing online forums, or communicating with other radio operators. Familiarity with these terms is vital; reading about two-way radios can feel like learning a foreign language (*Baofengradio*, 2021)!

Knowing the proper definitions helps you effectively configure and operate your radio, troubleshoot any issues, and research problems through online knowledge bases. Understanding their terminology allows you to tap into the broader radio communication community (*Amateur Radio Terms*, 2022).

Don't be overwhelmed by the jargon! Start with the most essential terms covered in this chapter. You will steadily pick up more terminology with practical experience. Refer to the glossary for help.

TUNING IN: THE BASICS OF FREQUENCY AND CHANNEL

Two key terms you'll continually encounter are frequency and channel. When operating your radio, understanding the distinction between entering a frequency versus selecting a channel is important.

In your Baofeng radio, frequency is the specific radio wave it sends out and picks up, akin to choosing a particular station on your car radio. It's quantified in Hertz (Hz), which counts the number of wave cycles occurring every second. Essentially, a greater frequency value corresponds to a higher frequency of the radio wave, much like tuning to different stations on your radio involves different frequency numbers.

Channels let you save groups of related frequencies under shortcut numbers for quick recall. For example, you may have channel 1 set to 154.570 MHz. So, when you select channel 1 on your radio, it automatically tunes to that saved frequency. Channels are an easy way to store and access the frequencies you use often. It's like setting your favorite music channels in your car for quick and easy access.

Baofeng radios can function in either frequency mode or channel mode (*General Mobile Radio Service*, 2023):

- **Frequency mode:** You directly enter the desired frequency using the numeric keys to access any frequency within the radio's range. In addition, if you enable the scanning feature, your radio will scan through each frequency.
- **Channel mode**: You select from pre-programmed channels configured with specific frequencies. This is quicker but restricts you to only those already set-up channels. When enabling scanning, your radio will scan through only your preset channels.

MAKING CONNECTIONS: THE ROLE OF REPEATERS

Repeaters play a vital role in two-way radio communications. A repeater is a device that receives radio signals on one frequency and retransmits them at a higher power on another frequency. This allows the signal to cover longer distances by boosting the radio signal strength. Amateur radio operators typically use repeaters to extend the range of their VHF and UHF band radios. For effective repeater operation, you need to understand terms like input frequency, output frequency, and offset (*Ham Radio Licenses*, n.d.):

- **Input frequency**: This is the specific frequency on which your Baofeng radio transmits to communicate with a repeater. When setting up your radio to use a repeater, you'll need to know this frequency to ensure your radio is sending signals that the repeater can receive. It's like dialing the correct number to make a phone call – you need to use the right frequency so the repeater can 'answer' your radio's call.

- **Output frequency**: Once your signal reaches the repeater, it doesn't just relay your exact signal. Instead, it rebroadcasts, or 'retransmits,' your signal on a different frequency, known as the output frequency. This is the frequency other radios in the area will tune into to hear your transmission. Using a different output frequency allows for efficient and clear communication, as it prevents your transmitted signal from directly interfering with the signal being received by others.
- **Offset**: The offset is a critical concept in repeater operation. It's the difference between the input and output frequencies. For example, if a repeater has an input frequency of 145.270 MHz and an output frequency of 145.870 MHz, the offset is 600 kHz. This separation is essential to avoid interference; it ensures that the repeater's transmissions don't interfere with the signals it's trying to receive. Think of it as a conversation where one person speaks at a time to avoid confusion – the offset allows the repeater to 'speak' and 'listen' at different 'tones,' avoiding a conversational clash.

When programming your Baofeng with a repeater, you must input the proper transmit and receive frequencies with the correct offset between them. Getting this configuration right is essential for making repeaters work smoothly.

In a later chapter, we'll provide step-by-step instructions on how to program frequencies and offsets for repeaters. For now, remember that repeaters operate using separate frequencies for receiving signals from your radio and retransmitting them.

SPEAKING THE SAME LANGUAGE: CTCSS AND DCS

CTCSS (Continuous Tone-Coded Squelch System) and DCS (Digital Code Squelch) are sophisticated systems used in ham radio to ensure clearer and more exclusive communications. These systems are particularly valuable in filtering out unwanted conversations on busy frequencies, acting like a selective hearing mechanism for your radio.

- **CTCSS**: Imagine being in a crowded room where everyone is talking, but you only want to hear conversations from your friends. CTCSS works similarly. It embeds one of 50 low-frequency tones into your transmitted audio signal. These tones are like secret handshakes; they are inaudible during transmission but are essential for communication. When you set a specific CTCSS tone on your Baofeng radio, it 'listens' exclusively for signals with that tone. This means your radio will remain silent (or squelched) until it detects a transmission with the matching tone, effectively filtering out all other chatter on the frequency.
- **DCS**: DCS is a more advanced version of CTCSS, using one of 105 digital codes instead of tones. Think of it as a digital password added invisibly to your communication. When you configure a specific DCS code on your radio, it acts as a gatekeeper, only allowing transmissions with the correct digital code to be heard. This is particularly useful in environments with a lot of radio traffic, as it ensures you receive only relevant communications.

In summary, CTCSS and DCS are integral features of your Baofeng radio, enabling you to participate in clutter-free and focused radio communications. By setting up these systems, you can ensure that your radio conversations are exclusive to your intended group, eliminating irrelevant transmissions from others on the same channel. Once configured in your radio's menu, this automated process enhances your communication experience by providing a more tailored, interference-free listening environment.

MODES OF COMMUNICATION: SIMPLEX AND DUPLEX

Two fundamental modes of communication in ham radio are simplex and duplex. Understanding the distinctions and applications of these modes is crucial for any amateur radio enthusiast.

- **Simplex Communication:** This mode is akin to a one-lane bridge, allowing two-way communication but only one direction at a time. For example, if a frequency of 146.520 MHz is set on a simplex channel, your Baofeng radio uses the same frequency to transmit and receive signals. It's ideal for straightforward, short-range conversations, as seen in walkie-talkies and basic ham radio setups. The main limitation is the inability to talk and listen simultaneously.
- **Duplex Communication:** Duplex, on the other hand, is like a two-lane bridge, enabling simultaneous two-way communication. For example, your radio might receive signals at 146.520 MHz but transmit at 147.120 MHz. Duplex can be either full (simultaneous talking and listening) or half-duplex (both functions, but not at the same time).

In essence, simplex and duplex are the building blocks of radio communication in ham radio. Simplex, with its one-at-a-time communication style, is straightforward and effective for short-range conversations. Duplex offers simultaneous interaction, which is more complex but essential for continuous dialogue. These modes shape how ham radio operators connect, each serving distinct purposes and scenarios.

LISTENING CAREFULLY: SQUELCH AND VOX

Two crucial features in Baofeng radios that significantly enhance audio reception and transmission are squelch and VOX (Voice-Operated Transmission). Understanding and properly configuring these functions can dramatically improve your radio communication experience.

- **Squelch:** This function acts as an audio gatekeeper. It mutes the speaker output when no signal is present, eliminating background static noise. The squelch threshold is adjustable to suit the RF noise levels in your environment. Properly setting the squelch level filters out weak signals, allowing only strong and clear communications to be heard. If set too low, it lets in too much background noise; set too high, and it might block desired signals. Thus, finding the right balance in squelch settings ensures crisp and uninterrupted reception.
- **VOX:** VOX automates the transmission process by detecting your speech and automatically triggering the transmit mode. This hands-free operation is beneficial when pressing the PTT (Push-To-Talk) button is inconvenient, like handling equipment or climbing. VOX sensitivity should be adjusted to activate with your voice while reliably avoiding unintended transmissions. This feature allows for smooth, natural conversation flow in radio communications without manual switching.

In summary, mastering the squelch and VOX features on your Baofeng radio can dramatically improve your communication experience. Squelch eliminates unnecessary background noise, ensuring that only relevant transmissions are heard. VOX, on the other hand, offers convenient hands-free operation, allowing for seamless conversations. Both features ensure efficient, straightforward, and user-friendly radio operation when optimally configured, making them indispensable tools for any Baofeng radio operator.

STAYING WITHIN BOUNDS: UNDERSTANDING RF GAIN AND BANDWIDTH

In the world of ham radio, two critical settings that can significantly impact the performance of your Baofeng radio are RF Gain and Bandwidth. Mastering these settings can significantly enhance your radio's range and audio quality.

- **RF Gain:** This controls how sensitive your radio is to incoming signals (Noonan., n.d.). Increasing RF Gain is akin to increasing the volume but for signal strength, not audio. It strengthens weaker signals and makes them more intelligible. It's beneficial for capturing distant or faint transmissions. However, too much RF Gain on strong signals can cause distortion. The key is to start with a moderate level and adjust upwards only as needed to improve the reception of weaker signals.
- **Bandwidth:** This refers to the width of the frequency range your radio can receive or transmit at any given time (*Choosing an Antenna*, n.d.). Wider bandwidth allows your radio to monitor a broader range of frequencies simultaneously, which helps scan busy bands. However, a narrower bandwidth provides better filtering from interference, focusing on a tighter frequency range. Choosing the right bandwidth depends on your situation: wider for monitoring more activity, narrower for clearer, more focused communication.

Understanding and correctly adjusting RF Gain and Bandwidth can significantly enhance your Baofeng radio experience. RF Gain helps you fine-tune the radio's sensitivity to pick up distant or weak signals, while Bandwidth allows you to control the range of frequencies you can access at once. Balancing these settings optimally will enable you to maximize your radio's performance, whether trying to reach far-off stations or need clarity amidst a crowded frequency band. These settings are vital tools in the ham radio operator's kit, enabling more effective communication.

SAFETY FEATURES: TIME-OUT TIMER (TRANSMISSION TIMER) AND EMERGENCY ALERT

Baofeng radios have essential safety features like the Time-Out Timer and the Emergency Alert, designed to enhance the user's safety and the radio's functionality. Understanding and utilizing these features can be crucial in various situations.

- **Time-Out Timer (TOT):** Also listed as 'Transmission Timer' in user manuals, it is a built-in safety mechanism that prevents the radio from transmitting too long. The timer typically has a preset limit, often around 60 seconds (*How Far Can I Talk?*, n.d.). If you transmit continuously past this duration, the Time-Out Timer automatically stops the transmission. This is particularly useful in preventing potential damage to the radio, such as overheating, which could occur from an accidentally pressed or stuck microphone button. The TOT is a safeguard, ensuring that transmissions are brief and the radio remains functional. To adjust the TOT:

  - Press the MENU key, followed by '9' on the keypad. This takes you to the 'TOT' (Time-out Timer) setting. Press MENU to enter.
  - Use the UP and DOWN keys to set the desired timeout duration. This is the maximum length of time you can transmit continuously. It will range from Off to 600 seconds.
  - Press the MENU key again to save your selection.
  - Hit EXIT to leave the menu. Your timeout timer is now set.

- **Emergency Alert:** This feature is a critical safety tool in emergencies. The radio triggers a loud siren and activates flashing LEDs by pressing a designated button, usually labeled "Alarm" or the 'Call' button (*How to Use Baofeng Radios*, 2022). This function is designed to attract attention in emergencies, especially when the user is in danger but unable to speak. The alert broadcasts a siren to nearby radios and provides a visual signal, making it an effective way to summon help. It's a vital feature for public safety, with personnel often trained to respond to these alerts. However, using this function responsibly is essential to avoid false alarms, which can undermine its effectiveness in real emergencies.

In conclusion, the Time-Out Timer and Emergency Alert are invaluable features in Baofeng radios. The Time-Out Timer ensures safe and responsible use by limiting continuous transmission and protecting the radio from damage. The Emergency Alert function provides a critical means of signaling for help in dangerous situations. Both features enhance the safety and utility of Baofeng radios, making them more reliable in various scenarios.

EXPANDING YOUR REACH: HAM RADIO AND DUAL BAND

Baofeng radios offer advanced features like ham radio operation and dual-band functionality, significantly enhancing their versatility for amateur radio enthusiasts. Understanding these features and their implications is crucial for anyone looking to maximize their Baofeng radio's capabilities.

- **Ham Radio Operation:** Refers to amateur radio, a hobby and service enabling long-range wireless communication and experimentation. You need a valid FCC-issued amateur radio license to access certain frequency bands and higher power operations (*Ham Radio Licenses*, n.d.). Baofeng radios cover various ham bands, so having at least a Technician-level license opens up a wider range of operating privileges. It's important to remember to only transmit on ham bands for which you're licensed.

- **Dual-Band Functionality:** A dual-band Baofeng radio can operate on two different frequency bands, typically VHF and UHF. This feature is like having two radios in one, allowing you to monitor a local UHF repeater while also scanning a VHF channel, for instance (*How Far Can I Talk?*, n.d.). Dual-band functionality, especially with the dual watch feature, enables simultaneous monitoring of two channels, expanding your communication possibilities.

While a ham radio license is required to fully utilize certain frequencies, Baofeng radios' dual-band capability makes them highly versatile, even for beginners. Understanding and using these features can significantly enhance your radio experience.

Venturing into the world of radio communications introduces many new terms and concepts. It's natural to feel a bit overwhelmed initially, but don't worry—gradually familiarizing yourself with these concepts is vital. The glossary section is a handy resource whenever you need to clarify terms or refresh your memory. It's designed to make your journey into radio communications smoother and more enjoyable.

With a solid grasp of the basic radio terminology in your toolkit, we're poised to dive into the exciting next chapter: programming your Baofeng radio. This is where the real adventure begins, as you unlock the full potential of your radio by tailoring it to your specific needs. From installing necessary software to inputting channels, we'll guide you through every step of the process, ensuring a beginner-friendly experience. You'll soon discover the impressive range of advanced features your Baofeng radio offers. So, let's roll up our sleeves and delve into the hands-on experience of radio programming!

4

BAOFENG PREPAREDNESS - MASTERING RADIO PROGRAMMING

A Baofeng radio isn't just a gadget; it's a lifeline. Whether traversing rugged landscapes, coordinating community safety initiatives, or keeping in touch during emergencies, the ability to program your Baofeng radio is a crucial skill in your preparedness toolkit.

Programming your radio equips you with the agility to adapt to changing situations, ensuring that you have quick access to essential frequencies, from local emergency channels to weather alerts. This chapter goes beyond the basics of radio use; it's about empowering you with the knowledge to tailor your Baofeng to your unique needs.

Whether manually programming channels or using software like CHIRP, these guidelines are tailored to help you maintain a crucial edge in readiness. You'll learn to navigate the radio spectrum with ease, ensuring that in any scenario – be it adventure, community service, or emergency – your Baofeng radio is an asset you can rely on.

Demystifying the Programming Process

Programming a complex communications device like a Baofeng radio can seem intimidating for a beginner. However, it simply involves entering frequencies you want to use into the radio's memory. This allows you to quickly tune in to programmed channels instead of manually entering a frequency each time (Centers, 2022).

Programming is like creating a speed-dial phonebook on your radio. You save significant time and effort by saving your most-used frequencies as channel entries rather than manually inputting them repeatedly.

Software programming requires a computer and cable to interface with your Baofeng and configure the settings. However, you can also manually input frequencies directly on the radio. This chapter will cover both methods.

MANUAL PROGRAMMING: THE OLD-SCHOOL APPROACH

Manually programming your Baofeng involves directly inputting frequencies using the numeric keypad. Let's walk through it step-by-step with a real-world example:

1. Find your local NOAA weather radio frequency:

- Go to www.weather.gov/nwr and click the "Find My Station" button.
- Enter your zip code into the 'NWR Station Search' section and press 'Enter.'
- Note the frequency in the yellow box. In my area, the frequency is 162.550 MHz.

2. Determine what memory channel you want to store this on.

Deleting a Saved Channel:

- If your radio came programmed with channels already saved, you may need to delete a saved channel to open it up for your use. To delete a channel on a UV-5R:
- Press 'MENU'.
- Enter '28' or use the arrows to find DELCH (delete the memory channel)
- Press 'MENU' to confirm and enter.
- Find the channel you want to delete and free up. Let's use channel #5 in this example. Arrow up to channel 5.
- Press 'MENU' to confirm and delete the saved channel.
- We now have a free channel to use.
- Press 'EXIT'.

Voice Prompts:

3. Press the VFO/MR button to enter frequency mode.

- To hear if you are in channel or frequency mode (and hear other operations), turn on 'Voice Prompts.'
- Press 'MENU'.
- Enter '14' or use the arrows to find VOICE (Voice Prompt).
- Press 'MENU.'
- Use the up arrow to switch to ENG (English) or CHI (Chinese). I assume you will want English (ENG) and then press 'MENU.' I will say it's fun to change it to Chinese and then scroll through all the menu options to hear Chinese!
- Press 'EXIT.'

4. Again, press VFO/MR; you should hear if you are on Channel or Frequency Mode. Ensure you are in frequency mode.

5. Use the numeric keys to enter the NOAA frequency in MHz format.

- For example, in my area, I will enter the frequency as 162.550 MHz and enter 162550 on the keypad.

Save a Channel

6. To save this frequency into a memory channel, press 'MENU.'

- Enter '27' or scroll up to MEMCH (stored in the memory channel)
- Press 'MENU.'
- Use arrows to go up to channel 5.
- Press 'MENU'
- You have now saved this frequency to channel 5.
- Press 'EXIT.'
- To test this out, press VFO/MR until you are on Channel Mode
- Use the arrows to select channel 5.
- Is the correct frequency displayed?

Always double-check your frequencies to avoid typos. Manual input takes patience, but it works in emergencies if your computer software is unavailable or you are deep in the woods. Manual programming will serve you well if you are starting with radio. However, as the years pass and your list of frequencies expands, upgrading to software is a significant time saver.

MANUAL PROGRAMMING: REPEATERS

Programming a repeater into your Baofeng can be a bit complex for beginners, but you can do it successfully with step-by-step instructions. Here's how to manually program a repeater frequency into your Baofeng UV-5R:

Step-by-Step Manual Programming

1. Switch to Frequency Mode.

- Turn on the radio.
- Press the 'VFO/MR' button to switch to Frequency Mode (VFO).

2. Enter the Repeater's Frequency.

- Use the keypad to enter the repeater's frequency. For example, if the repeater's frequency is 146.940 MHz, type 146940.

3. Set the Transmit Offset.

- The standard offset for a 2-meter band VHF (144-148 MHz) repeater is usually 600 kHz, while it is often 5 MHz for a 70 cm band UHF (420-450 MHz) repeater.
- Determine if the offset is positive or negative based on the repeater specifications.
- Your repeater resources will typically give you both of these.

4. Set the Offset Direction and Amount on your radio.

- Press the 'MENU' button.
- Enter **25** (for frequency direction) SFT-D, and press 'MENU.'
- Use the up/down arrow keys to select the offset direction (N for simplex, + for positive, - for negative). For most repeaters, you'll select + or -.
- Press 'MENU' to confirm.
- Press 'EXIT' to return to the main screen.
- Press 'MENU' again.
- Enter '26' (for offset frequency) OFFSET, and press 'MENU.'
- Enter the offset amount (e.g., 000600 for 600 kHz) using the keypad.

- Press 'MENU' to confirm.
- Press 'EXIT' to return.

5. Set the CTCSS/DCS Code (If Required)

- If the repeater requires a CTCSS/DCS tone for access, you must set it. Your repeater resource will provide this information.
- Press 'MENU'.
- Enter '13' (for CTCSS) T-CTCS or '12' (for DCS) T-DCS, and press 'MENU.'
- Use the keypad or the up/down arrows to select the correct tone.
- Press 'MENU' to confirm.
- Press 'EXIT' to return.

6. Save the Channel following the instructions from the manual programming section.

Final Steps

- Turn the radio off and then back on.
- Press the VFO/MR button to switch to Channel Mode.
- Use the up/down arrows to find your new channel.

To verify if you have correctly set up your Baofeng to communicate with a repeater, you can follow these steps:

Check Your Settings:

- Ensure the frequency and the offset are correctly entered.
- Confirm that the offset direction (plus or minus) is set according to the repeater's requirements.
- Verify the CTCSS/DCS tone is correctly set if the repeater uses one.

Listen to the Repeater:

- Before transmitting, listen to the repeater's frequency to see if you can hear other conversations. This can confirm whether you're on the right frequency and whether the repeater is within range.

Perform a Radio Check:

- When the frequency is clear, key up (press the transmit button) and briefly state your call sign along with a request for a radio check. For example, "This is [Your Call Sign] requesting a radio check."
- If your setup is correct and someone is listening, you might receive a response.
- Only do this step if you are legally licensed for this frequency.

Check for a Repeater's Automated Response:

- Some repeaters have an automated voice or a beep tone that indicates you have successfully accessed the repeater.
- After transmitting, if you hear this automated response, it's a good sign that you're set up correctly.

Ask for Confirmation from Local Hams:

- If you're part of a local ham radio club or have ham friends nearby, ask them to listen and confirm if they can hear you through the repeater.

Check the Transmission Quality:

- If you do make contact with someone, ask for feedback on the quality of your transmission. They can tell you if your signal is clear, too weak, or distorted.

Adjust Antenna or Location if Necessary:

- If you cannot hit the repeater or receive poor reports on your signal, consider adjusting your antenna or moving to a location with less obstruction.

Use a Different Radio to Monitor:

- If you can access another receiver or a different radio, you can listen to the repeater's frequency while transmitting from your Baofeng. This can help you hear what your transmission sounds like.

Remember, success in accessing a repeater can depend on various factors, such as the power output of your radio, your distance from the repeater, the quality of your antenna, and geographical obstructions.

SOFTWARE PROGRAMMING: PREPARING FOR THE PROCESS

Software programming allows you to configure your Baofeng radio's settings and memory channels using a computer. Let's break down the programming process using CHIRP, a popular open-source software program (Centers, 2022). Here are some tips to get prepared:

1. Download programming software:

- Go to https://chirp.danplanet.com/projects/chirp/ wiki/Home and click "Download CHIRP for your platform."

  - You can also go to Google and search for "Chirp danplanet," and it will be the first thing that comes up.
  - Warning: there are ads on this site that have a

'Download' button. DO NOT click on those! Click only in the text or paragraphs where it says 'Download.'

- Select the version for your operating system. CHIRP is free and works with most Baofengs. The version for your computer will usually be highlighted in green.
- Follow your computer prompts to download the software.

2. Install CHIRP on your computer following standard software installation steps. Ensure you download the software version compatible with your specific Baofeng model. Then, open the program.

3. Connect the USB programming cable from your radio to your computer.

- If you are like me, your programming cable came with a CD to install the driver. Do this first so that you install the driver from the CD. Before I did this, CHIRP didn't recognize my cable or radio.
- Note: CHIRP will ask you what computer port your cable is connected to. If you are like me, you may not know, so I always choose the 'HELP ME' option in the drop-down and follow the instructions.

4. You should have obtained a list of frequencies you want to program.

- Search databases like RadioReference.com to find local repeater and public safety frequencies.
- Refer to your local radio club's recommended frequencies.

5. Consider your intended use—emergency, local communications, private channels, etc.- as this will determine your channels, settings, and how you want to group them.

- Pro-Tip: Organize your thoughts on paper or a spreadsheet. Group things out here before doing the work in the software.

These steps will help ensure a smooth programming experience using CHIRP.

PROGRAMMING WITH SOFTWARE (CHIRP) IN MEMORY EDITOR

1. Launch CHIRP.

2. Connect your Baofeng radio to your computer using the supplied programming cable.

3. In CHIRP, go to the 'Radio' menu at the top of the screen and click 'Download from Radio' to download your current radio memory contents as a .csv file. Use 'Help Me' to find your port.

4. Save this to your computer as a backup copy.

5. Your memory editor will display and look like the image below.

CHIRP (Baofeng_UV-5R_20240115.img)

File Edit View Radio Help

Baofeng_UV-5R_20240115.img* ×

Memories Settings

	Frequency	Name	Tone Mode	Tone	Tone Squelch	DTCS	RX DTCS	DTCS Polarity	Cross mode	Duplex	Offset	Mode	Skip	Power	Comment
0	162.550000														
1	162.550000											FM		High	
2															
3	454.325000		TSQL		136.5							FM		High	
4	455.425000		Cross		151.4				->Tone	split	162.550000	FM		High	
5	456.525000		TSQL		192.8							FM		High	
6	457.625000		TSQL		241.8							FM		High	
7	458.725000		DTCS			025		NN				FM		High	
8	459.825000		DTCS			134		NN				FM		High	
9	461.925000		DTCS			274		NN				FM		High	
10	462.225000		DTCS			346		NN				FM		High	
11	463.325000		DTCS			503		NN				FM		High	
12	464.425000		DTCS			073		RR				FM		High	
13	465.525000		DTCS			703		RR				FM		High	
14	402.225000											FM		High	
15	437.425000											FM		High	
16	479.975000											FM		High	
17	138.550000											FM		High	
18	157.650000											FM		High	
19	172.750000											FM		High	
20	438.500000											FM		High	
21	155.700000											FM		High	
22															
23															

Baofeng UV-5R

6. Enter your desired frequencies and settings in the Memory channel slots.

7. Once your channels are configured, click the 'Radio' menu and select 'Upload to Radio' to program your entered settings and frequencies into the radio.

8. Remove the cable to test. Toggle through the channel numbers on your radio to verify your programmed frequencies and settings.

Understanding the CHIRP Fields

Location: The far-left-hand column is filled with numbers. This is the channel number or memory location in your radio, and your radio's capacity defines the number you see.

Frequency: Sets the channel's receiving frequency in megahertz (MHz). If Duplex is set to None, it also serves as the transmit frequency.

Name: An alphanumeric label for the channel displayed on the radio's front panel. Your radio's display capabilities define the character limit.

Tone Mode: Controls transmit/receive squelch tones or selective calling features.

- Options:

  - None: No tone or code transmitted; receive squelch is open or carrier-triggered.
  - Tone: Transmits a single CTCSS tone; receive squelch is open or carrier-triggered.
  - TSQL: Transmits and requires a matching CTCSS tone for receive squelch.
  - DTCS: Transmits a DTCS/DCS code; receive squelch requires the same code.

○ Cross: A combination of squelch technologies; specifics set in the Cross Mode field.

Tone: Sets the CTCSS tone for transmission when Tone Mode is set to Tone or Cross.

Tone Squelch: Sets the CTCSS tone for transmission and receive squelch in TSQL or Cross Mode.

DTCS: Sets the DTCS code for transmission and receive squelch in DTCS or Cross Mode.

RX DTCS: Sets the DTCS code for receive squelch when using Cross Mode (if supported by the radio).

DTCS Polarity: Determines DTCS code polarity for both transmitting and receiving.

- Options:

  ○ 'N' for Normal
  ○ 'R' for Reversed.

Cross Mode: Configures complex squelch behaviors when Tone Mode is set to Cross. Consists of two technologies separated by an arrow (->). The left side controls the transmit selective-call method, and the right side controls the receive squelch.

- Possible Values:

  ○ Tone: Uses CTCSS tones (transmit tone from the Tone column, receive tone from the ToneSql column).
  ○ DTCS: Uses DTCS/DCS codes (transmit code from the DTCS Code column, receive code from the DTCS Rx Code column).
  ○ Combination: For example, "Tone->DTCS" means a

CTCSS tone is used for transmission (set in Tone), and a DTCS Rx Code is used for receive squelch.

- Examples of Cross Mode configurations include "Tone->Tone," "DTCS->Tone," and "->DTCS," each combining different technologies for transmit and receive squelch functions.

Duplex: Sets the channel's duplex mode.

- Options:

  - None: Transmit and receive frequencies are the same.
  - +/- (Plus/Minus): Transmit frequency is above/below the receive frequency by the Offset value.
  - Split: Uses an absolute transmit frequency, as defined in Offset.
  - Off: Disables transmission on the channel.

Offset: Defines the frequency difference for transmit in relation to receive frequency (if Duplex is not None or Off). Enter the shift amount (in MHz) or absolute transmit frequency for Split Duplex.

Mode: Controls the channel's transmit and receive mode.

- Common Modes:

  - FM: Wide FM (5 kHz deviation) for two-way communications.
  - NFM: Narrow FM (2.5 kHz deviation) for two-way communications.
  - WFM: Wide FM for broadcast (approx. 100 kHz deviation).

- ○ AM: Used in the aircraft band in the US.
- ○ DV/DN: Digital modes for specific systems (e.g., D-STAR, SystemFusion).

Skip: Controls scan behavior for the channel.

- • Options:

  - ○ S: Skip this channel during a scan.
  - ○ P: This channel is a priority.

Power: Sets the power output for this channel.

Comment: Space to enter your notes about this channel.

PROGRAMMING WITH SOFTWARE (CHIRP) WITH CSV FILES

CSV stands for "Comma Separated Values," a flexible data encoding format rather than a strict file format.

Creating and Editing a CHIRP-Compatible CSV File

Starting from Scratch:

1. Start CHIRP: Open the CHIRP software.
2. Create a New File: Navigate to File > New to generate an empty CSV file.
3. Add a Memory: Input a frequency (e.g., 155.525) in the first row to create a memory entry.
4. Export the Template: Save your new file with a .csv extension via File > Export.
5. Edit in Spreadsheet Software: Open this CSV file in a spreadsheet application, make your edits, and save.

6. Reopen in CHIRP: To check the format, open the edited file in CHIRP by going to File > Open.
7. Copy Memories: Optionally, you can copy these memories to another radio tab within CHIRP.

Tips for Maintaining Formatting

CHIRP depends on the CSV's format to correctly read and process the data. Follow these guidelines to avoid issues:

- Preserve the Header Row: Keep the original header format so CHIRP knows the value of each column.
- Reorder with Care: You can change the column order if the header row remains consistent.
- Deleting Columns: You can remove columns with default values; CHIRP will assume these defaults when opened.
- Quoting Values: Use double quotes (") around values containing spaces or commas.
- Field Value Accuracy: Ensure field values match the original and accepted formats. Refer to a CHIRP-generated CSV file for examples of correct values.

Transferring Data to Your Radio

Since CSV files are generic and not specific to any radio model, direct uploads to a radio are not possible. Follow these steps to transfer data:

1. Open CSV File: Use File > Open in CHIRP to open your CSV file.

2. Load Radio Image: Open a .img file for your radio, or download one from your radio using Radio > Download From Radio. Ensure the tab indicates the specific radio model.

3. Select and Copy Memories:

- Select the tab with your CSV file.

- Highlight the memories you wish to transfer (use Shift or Control for multiple selections).
- Copy them using Edit > Copy.

4. Paste into Radio's Memory:

- Switch to the tab for your actual radio.
- Choose the memory slot to paste into, then use Edit > Paste.

5. Uploading Changes:

- For clone-mode radios, upload the changes back to your device with Radio > Upload To Radio.
- For live-mode radios, wait for synchronization to complete (check the status bar).

Handling Missing Columns

If the CSV lacks information needed by your radio (like a "Power" column), CHIRP can still help:

1. Adjust in CHIRP Radio Window: Open the spreadsheet in CHIRP's radio-specific window.

2. Bulk Edit Rows:

- Highlight relevant rows (click the first row, shift-click the last).
- Right-click and choose "properties."
- Check the property ("column") you wish to change and select the new value from the dropdown (e.g., "power" to "high").
- Click OK to apply changes.

Following these steps and guidelines, you can efficiently use CHIRP to program your radio, ensuring data is formatted and transferred correctly.

I also realize this is a lot of new information. Return to this section until you are 100% comfortable completing these actions.

OVERCOMING COMMON PROGRAMMING CHALLENGES

Despite the step-by-step process, you may face some common challenges when programming your Baofeng radio:

- Compatibility issues if your software version needs to be updated. Check CHIRP's website and upgrade to the latest release.
- Errors in channel numbers or frequencies cause incorrect programming. Double-check your inputs and file contents. Make sure you have all your decimals right!
- Failure to connect radio with computer. Try different USB ports and cables. Enable USB debugging mode on the radio. Here are some steps to debug and troubleshoot if you are unable to connect your Baofeng radio to your computer:

  - Try different USB ports on your computer.
  - Test different USB cables. Cables can fail over time.
  - Restart the CHIRP programming software on your computer and reconnect the cable. Sometimes restarting can clear up USB issues.
  - Update USB drivers for your computer's operating system and Baofeng model. Outdated drivers can disrupt connectivity.
  - As a last resort, try programming software that doesn't require a cable.

- Software freezing or glitching. Reinstall your software and upgrade computer drivers. Restart your PC.
- Overwriting previous program memories accidentally. Always back up your radio memory before programming.

- Loose antenna or battery interrupting the programming process. Ensure all connections are tight.

Remember, encountering a roadblock now and then is part of learning new skills. By troubleshooting these common programming issues, you'll resolve the problem and gain a deeper understanding of how your Baofeng radio works. So, take your time, and don't get frustrated with initial hiccups. Programming gets much easier with practice. **Always back up and verify your inputs**.

————

Now that you have a programmed Baofeng radio customized to your needs and interests, it's time to focus on achieving the best possible voice communications with your equipment. In the next chapter, we'll explore tips and techniques to make your transmissions crystal clear every time you press that push-to-talk button. From antenna optimization to mic technique and cutting through noise, you'll learn how to make your radio's audio as clean and understandable as possible. Let's move on to ensuring your team can perfectly hear every word coming through loud and clear!

5

CRYSTAL CLEAR – SECURING COMMUNICATIONS

I n the world of radio, where preparedness and clear communication can mean the difference between safety and peril, your Baofeng radio is a vital tool. This chapter is dedicated to honing your skills in achieving crystal-clear radio transmissions, an essential aspect for any prepper.

Here, we dive into the nuances of signal clarity, exploring methods to ensure pristine reception and troubleshooting techniques to eliminate interference. Not only will you learn to banish noise but also to amplify your radio's capabilities through repeater use and optimal antenna selection. This knowledge is invaluable, especially in emergencies when reliable communication can significantly impact your survival strategy.

Get ready to elevate your Baofeng radio from a basic communication device to an exceptional tool in your preparedness arsenal, ensuring it won't disappoint you when you need it the most.

GETTING STARTED: BAND SELECTION

Understanding the nuances of VHF (Very High Frequency) and UHF (Ultra High Frequency) bands is crucial for clear and effective communication. Each band has its unique characteristics and is suited for different environments.

VHF (Very High Frequency) Bands:

- Remember, 'V' for very long range.
- Frequency Range: 136–174 MHz.
- VHF signals excel in open outdoor areas, providing extended range due to line-of-sight propagation. They travel in straight lines, making them suitable for distances without significant physical barriers. A 5-watt VHF signal can optimally reach 1–3 miles (Centers, 2022).
- Indoors, VHF's performance is limited (around 100–500 feet) because its longer wavelengths struggle to penetrate walls and other obstacles.

UHF (Ultra High Frequency) Bands:

- Remember, 'U' is for urban, so better at penetrating obstacles.
- Frequency Range: 400–520 MHz.
- UHF signals are more effective in environments with obstructions like buildings and foliage. Their shorter wavelengths allow better penetration through barriers, offering improved indoor coverage.
- In open outdoor settings, UHF has a shorter range (approximately 0.5–1 mile) than VHF, but its ability to navigate obstructions can be advantageous in urban or forested areas.

Understanding these bands helps you select the right one based on your operating environment. Additionally, tweaking other radio settings, such as squelch, bandwidth, and RF gain, and using the correct antenna height and orientation can further enhance signal quality. The frequency band and the radio's output power heavily influence the transmission range.

In summary, grasping the science behind VHF and UHF propagation is essential to realistic expectations regarding range and signal quality (Amateur Radio Service, 2016). Your Baofeng radio supports both bands, allowing you to choose the best frequency for your specific environment, whether communicating in open fields or navigating the challenges of urban or indoor settings. By understanding and leveraging the unique characteristics of VHF and UHF bands, you'll be better equipped to maximize your radio's capabilities and maintain uninterrupted communication even during disasters.

SQUELCH (SQL)

Mastering the squelch function on your Baofeng radio is a crucial skill for any ham radio enthusiast. Squelch is essentially a gatekeeper for your radio's speaker output, determining when to mute unnecessary background noise and when to let through actual transmissions. A well-adjusted squelch setting is vital for a pleasant and efficient radio experience.

Understanding Squelch Levels:

- Squelch settings typically range from fully open (low number) to very tight (high number). A lower squelch level (around 3) means the radio will pick up more sounds, including background static. A higher setting (like 8) filters out weaker signals, reducing noise but potentially missing faint transmissions.

Balancing Squelch Settings:

- The goal is to find a balance where the squelch mutes background static without cutting off legitimate communications. Too loose, and you'll constantly hear static; too tight, and you might miss important signals.

Using Carrier Squelch:

- If you're plagued by constant static, engage the carrier squelch. Set the squelch level to 0 (fully open) and press the [Moni/Carrier] button. This mode mutes the speaker unless a signal is present, helping to filter out the static hiss. Adjust upwards from level 0 until the noise is muted, but communications still come through clearly.

Advanced Squelch with CTCSS and DCS:

- For environments with persistent static, consider using tone squelch options like CTCSS or DCS. These settings add a layer of noise filtering, allowing only transmissions with specific tones to be heard.

Using squelch effectively is about tailoring your radio to your environment and needs. A tighter squelch or carrier squelch mode can be invaluable in noisy settings. But remember, the perfect squelch setting is a moving target, varying with your surroundings and the specific signals you're trying to catch.

By understanding and adeptly adjusting the squelch settings, you can significantly enhance your radio communication, making it clearer, more focused, and more enjoyable. Whether chatting with fellow ham operators or scanning during a disaster, a well-adjusted squelch is vital to clear communications.

TROUBLESHOOTING POOR SIGNAL RECEPTION

Troubleshooting poor reception is a crucial skill for ensuring clear and effective communication. The difference between spotty and crystal-clear reception often lies in understanding and addressing common issues. Here's a guide to help you diagnose and solve reception problems with your radio.

Location and Environmental Factors:

- Reception quality can be significantly affected by your physical location, weather conditions, and nearby structures. Moving to a higher elevation or away from large buildings can improve signal strength.

Antenna Issues:

- The antenna is a common culprit for poor reception. Check the antenna connectors for looseness or damage. Upgrading to a higher-gain antenna can make a substantial difference.

Frequency and Offset Settings:

- Incorrect frequency or offset settings can prevent your radio from receiving transmissions correctly. Double-check your frequency inputs and ensure they are accurate for your intended communication channel.

Squelch and Tone Settings:

- An improperly set squelch level can let in too much noise or block out legitimate signals. Adjust the squelch level to find the right balance. Also, verify CTCSS/DCS tone settings if you're using them, as incorrect tones can block desired signals.

Battery Power:

- Low battery power can lead to reduced transmit power and, thus, poorer reception. Regularly check your battery level, and keep it charged. Carrying spare charged batteries can be a lifesaver in prolonged situations.

Adjust the Bandwidth: The bandwidth setting on your Baofeng radio is like adjusting the focus on a camera lens. It helps you get a clear "picture" by filtering out unwanted "noise." Changing the bandwidth can improve signal clarity, especially in crowded or noisy environments.

- On a UV-5R, press MENU
- Press '5' on your keypad (or use the arrows). This will take you to the 'WN' (Wide / Narrow) Bandwidth setting. Press MENU to enter.
- Using the UP and DOWN keys, select either 'WIDE' or 'NARROW.'
- Press MENU to save your choice.
- Press EXIT to leave the menu. Your bandwidth is now set.
- Try both settings to determine what works best in your current situation.

Poor reception typically involves checking equipment settings, ensuring optimal physical setup, and maintaining your gear. By methodically checking these areas – antenna, location, frequency, squelch and tone settings, and battery power – you can often pinpoint and resolve the root cause of reception issues. Remember, even minor adjustments can lead to significant improvements in reception quality.

This troubleshooting process will improve your experience and deepen your understanding of radio operations, making you a more adept and versatile operator.

ADVANCED TACTIC: FIND A REPEATER

Utilizing local repeaters is a crucial strategy for extending the range of your communications. Repeaters, essentially radio relay stations, capture your transmission and rebroadcast it at a higher power over a wider area. Knowing how to find and use them effectively can significantly enhance your ham radio experience.

Finding Repeaters:

- Online Resources: Websites like RepeaterBook (www. repeaterbook.com) offer comprehensive listings of local repeaters. Use GPS coordinates for precise locations.
- Mobile Apps: Apps such as RepeaterBook provide convenient access to repeater information on the go, allowing you to find nearby repeaters easily.
- Local Radio Clubs: Joining local ham radio clubs or their social media groups can provide valuable information on area repeaters, updates, and advice from experienced operators.
- Field Experience: You'll naturally learn about local repeaters as you engage more in ham radio activities and converse with fellow operators.

Using Repeaters:

- Etiquette: Always monitor a repeater before transmitting to ensure it's not currently in use and avoid interrupting ongoing conversations.
- Programming Your Radio: Input the repeater's transmit and receive frequencies with the correct offset into your radio.
- Understanding and utilizing local repeaters is a valuable skill to extend your range. Repeaters extend the range of your communications, which can mean life or death in a crisis; remember our opening story about Alden Jones. Learning to

find and use repeaters correctly opens up a world of enhanced communication possibilities.

ADVANCED TACTIC: UPGRADE ANTENNAS

Enhancing your Baofeng radio's performance through antenna upgrades is vital to achieving better reception and a broader communication range. The right antenna can significantly amplify your radio's capabilities, whether you're on the move or stationed at home.

Handheld Antenna Upgrades:

- For portable handheld units, consider a high-gain rubber duck antenna. Options like the Nagoya NA-320 or Abbree folding antenna can offer 3–4 times the range of standard antennas, making them ideal for on-the-go use.
- High-gain antennas focus their energy in a particular direction instead of omnidirectional antennas that radiate or receive signals in all directions equally. By concentrating the signal in a specific direction, high-gain antennas effectively increase the signal's strength in that direction.

Mobile Antennas:

- For mobile operation, magnetic mount antennas are a great choice. They are easy to install and remove, with no permanent installation required. Dual-band 2m/70cm antennas from brands like Comet or Diamond are popular and effective.

Base Station Antennas:

- At a fixed location, roof-mounted base station antennas can provide excellent range in all directions. Opt for durable materials like fiberglass or steel, and choose dual-band antennas for versatility.

Important Considerations:

- Ensure your chosen antenna is well-matched to your radio's frequency range and output power.
- Use high-quality, low-loss feed line cables like RG-8X to minimize signal loss.

Upgrading Antennas:

- Yagi Antennas: Yagi antennas are directional antennas with multiple elements, including a driven element, reflector, and directors. They offer high gain and are widely used in VHF/UHF applications for focused, long-distance signal reception and transmission.
- J-Pole Antennas: The J-pole antenna is a simple, efficient, vertically polarized antenna offering modest gain. It is commonly used in VHF/UHF amateur radio for its omnidirectional pattern and ease of construction. Roll-up J-Pole antennas provide a portable option, which is excellent for field use.
- Discone Antennas: The discone antenna is a wideband, omnidirectional antenna with a unique disc-cone structure. It is primarily used for VHF/UHF frequencies and is valued for its broad bandwidth and non-directional radiation pattern.
- Ladder Antenna: A ladder line antenna, often used in amateur radio, is a balanced feeder with two parallel conductors connected by rungs, known for its low-loss transmission, especially effective for high-frequency bands.

Upgrading your Baofeng radio antenna is a strategic move towards better, more reliable communication. A better antenna can transform your communications, whether in the field, on the road, or at your base station. By carefully selecting and correctly installing an appropriate antenna, you can significantly extend your reach and achieve clearer transmissions.

In the next chapter, we will navigate the important legal aspects of radio operations, including obtaining licenses and call signs.

The Power of Communication

"He who is not prepared today, will be less so tomorrow."

— *OVID*

Remember the story of Alden Jones? It was that incredible story of survival that inspired my journey to discover the powerful role of radio in preparedness.

There was no doubt in my mind that Baofeng radio had played a pivotal role in Alden's ability to survive, and it became clear to me that radio absolutely had to be part of any thorough preparedness plan.

My own peace of mind has been significant since I discovered all I have about ham radio, and it's become very important to me to share what I know to help other people improve their chances of survival no matter what happens.

The main problem I encountered when I first delved into this world was the complexity of information available about radio communication. It's vital that devices like Baofeng radio are accessible to people regardless of whether they understand the jargon that surrounds it or not, and it's my drive to make sure this happens.

But there's one more piece to the puzzle: The information is here, but now I need to make sure that the people who are looking for it can find it ... and that's where you come in.

By leaving a review of this book on Amazon, you'll help others realize how important this tool is to survival and show them exactly where they can find the guidance they need to use it.

In just a few minutes, you can transform someone else's survival plan. It's that easy.

Thank you so much for your support. Community is vital to survival, and I'm glad to have you as part of this one.

6

LICENSES - STAYING LEGAL WITH BAOFENG

A dherence to legal guidelines is paramount in the radio realm, so it's crucial to understand the regulatory framework governing Baofeng radios. These radios offer affordable and efficient two-way communication, essential in emergency and survival scenarios. However, they operate within a legal framework set by federal regulations.

This chapter is designed to navigate the complex legal landscape surrounding the use of Baofeng radios. It provides clear guidance on using these powerful communication tools responsibly, ensuring you stay within legal boundaries and avoid penalties.

For preppers, this knowledge is not just about compliance; it's about ensuring uninterrupted legal communication capabilities in times of crisis. Understanding these regulations enhances your preparedness strategy, keeping you informed and compliant and ensuring your communication lines remain open and legal when you need them most.

Making Sense of Licensing Requirements

In the United States, the Federal Communications Commission (FCC) regulates the airwaves and designates which frequency bands require licenses for transmission. As a Baofeng radio owner, you must understand if a license is needed based on your intended use. Transmitting without a license on restricted frequencies can potentially lead to fines, equipment seizure, loss of privileges, and even criminal prosecution in egregious cases. Let's make sure you stay legal.

UNLICENSED FREQUENCY BANDS

Some frequency bands are available for public use without any license required. These are generally allocated for short-range personal and business communications. Common unlicensed bands used by Baofeng radios include (*Family Radio Service* [FRS], 2011; *General Mobile Radio Service [GMRS]*, n.d.):

- **Family Radio Service (FRS)**: Uses the UHF 462–467 MHz band and is license-free but requires FCC-approved radios transmitting at 2 watts or less.
- **Multi-Use Radio Service (MURS)**: This service uses 5 VHF frequency channels near 151–154 MHz. No license is needed, but power is limited to 2 watts.

As long as you operate on these unlicensed bands within the power and equipment rules, no FCC license is required to transmit with your Baofeng radio.

LICENSED RADIO SERVICES

Legal radio communication involves understanding the licensing requirements for different frequency bands. Here's a breakdown of the licenses needed for specific radio services (General Mobile Radio Service [GMRS], n.d.; Ham Radio Licenses, n.d.):

General Mobile Radio Service (GMRS) Licensing:

- A GMRS license allows you to operate high-power two-way voice communication on the 462/467 MHz band.
- This license, which costs $35 and is valid for five years, covers you and your immediate family members.
- Unlike ham radio, there's no exam required to obtain a GMRS license.

Amateur "Ham" Radio Licensing:

- Ham radio operators have access to a wide range of frequency bands, each with its own characteristics and uses.
- You must pass an FCC exam to obtain a valid operator's license.
- There are three license classes - Technician, General, and Extra - each offering increasing access and privileges.
- The difficulty of the exam and the privileges granted ascend with each class.

It's crucial to note that operating on restricted frequencies without a valid FCC license is illegal and subject to enforcement actions. This includes public safety, aircraft, maritime, and other regulated bands. Always ensure you have the appropriate license for your intended radio use to avoid legal complications and to operate within the bounds of FCC regulations. Understanding and adhering to these licensing requirements is critical to a responsible and lawful radio communication experience.

'The Command Packet' includes a printable version of the FRS and GMRS frequencies, which you should keep close.

www.MorseCodePublishing.com/Command

BAOFENG-SPECIFIC GUIDANCE

Operating a Baofeng radio involves understanding and adhering to specific guidelines to ensure legal and responsible use. Here's a concise guide on what you need to know:

Licensing for Ham Radio Bands:

- To transmit on amateur (ham) radio bands (access to all amateur radio frequencies above 30 MHz), you must have a technician-class or higher amateur radio license.

GMRS Band Usage:

- If you wish to use the GMRS band, especially at higher power levels, you need a GMRS license. This license is more straightforward to obtain as it doesn't require an exam.

Adhering to Power Limits:

- Be mindful of power limits on certain bands like FRS (Family Radio Service) and MURS (Multi-Use Radio Service). On your Baofeng radio, ensure the power output setting does not exceed the allowed limits, such as 2 watts on the FRS band.

Restrictions on Certain Bands:

- Transmitting on commercial, public safety, emergency services, or critical infrastructure bands is illegal without proper authorization. Always check and comply with the regulations for each band.

Monitoring vs. Transmitting:

- While it's legal to listen to any frequency band for monitoring purposes, transmitting on restricted bands without a license is prohibited.

Remember, while Baofeng radios can transmit across a broad spectrum of frequencies, you must restrict your usage to the bands you're legally allowed to operate on. An FCC license grants legal access to certain bands and often allows for higher power output than unlicensed bands. Consider obtaining an appropriate license to enhance your radio operating experience and be fully prepared.

FCC RADIO OPERATOR'S LICENSE FOR ULTIMATE PREPAREDNESS

Expanding your radio communication capabilities often means obtaining an FCC license, which unlocks access to additional frequencies and increased power limits. Here's a detailed guide to the different types of FCC licenses available:

General Mobile Radio Service (GMRS) License:

- Cost: $35 for a 5-year term.
- Family Coverage: The license covers immediate family members.
- No Exam Required: Easier to obtain as no exam is necessary.

- Frequency Access: Grants access to the 462/467 MHz band with higher power limits up to 50W.
- Repeater Use: Permits the use of repeaters and duplex operations.
- Ideal for short-range personal and family communication needs.
- You will receive a call sign specific to GMRS. Unlike amateur radio, where operators often choose call signs, the FCC automatically assigns GMRS call signs primarily for identification and regulatory purposes.

Amateur "Ham" Radio License:

- License Classes: Three classes - Technician, General, Extra, each with increasing privileges.

  - Technician License: Entry-level; basic access to specific HF, VHF, and UHF bands.
  - General License: Offers more extensive HF band privileges.
  - Extra License: Provides full access to all HF bands and privileges.

- Exams Required: Each class requires passing an exam, with difficulty increasing for each subsequent level.
- Study and Exam Process: Materials and exams are administered by Volunteer Examiner Coordinators.
- Call Sign Issuance: Upon licensing, you receive a call sign to operate on amateur bands.
- Perfect for radio hobbyists and those interested in technical learning.
- A user can apply for a new call sign after reaching each new level to indicate their new status or to keep their existing call sign.

Whether looking for short-range family communication with a GMRS license or diving deeper into amateur radio with a Ham license, these options provide greater freedom and capabilities in your radio use. The choice depends on your specific needs and interest in the hobby, with the Ham radio license offering a path for technical growth and a more comprehensive range of communication experiences.

Ham Radio Licenses Expanded:

In the United States, the Federal Communications Commission (FCC) oversees the licensing of amateur (ham) radio operators, offering three main types of licenses. Each license type caters to different levels of expertise and access within the ham radio spectrum:

1. Technician License:

- Level: Entry-level, ideal for beginners in ham radio.
- Exam: A 35-question test covering radio theory, regulations, and operating practices.
- Privileges: Access to all amateur radio frequencies above 30 MHz, including the popular 2-meter and 70-centimeter bands. Limited privileges on certain HF bands (10 meters), mainly for Morse code.
- Why You Want It: The entry-level license is ideal for beginners. It is excellent for local and regional communication via VHF/UHF and basic HF privileges. It is popular with those interested in local communication, emergency communication, and getting started with amateur radio.

2. General License:

- Level: Intermediate for those seeking broader communication capabilities.

- Exam: Another 35-question examination, building upon the Technician level.
- Privileges: Expanded HF band privileges and access to most amateur bands, including most of the HF spectrum, allowing for international and long-distance communication.
- Why You Want It: This class is for those who want to explore the full range of amateur radio, including long-distance (DX) contacts. The General class opens up a world of global communication on the HF bands, where much of amateur radio's magic lies.

3. Amateur Extra License:

- Level: Advanced for those pursuing the fullest extent of amateur radio privileges.
- Exam: A comprehensive 50-question examination.
- Privileges: All amateur band privileges. Access to exclusive frequency allocations in the HF bands.
- Why You Want It: It's the highest level of licensing, appealing to those deeply involved in the hobby. It grants maximum operational flexibility and access to all available frequencies and modes. It is ideal for the most serious hobbyists who want to explore every aspect of amateur radio, including those interested in the hobby's advanced technical and experimental aspects.

Each progressive level of licensing requires passing a more challenging examination, which tests knowledge of radio theory, regulations, and operating practices. The choice of license depends on your interests in the hobby:

- Local Communication & Basic HF: Technician.
- Global HF Communication & Advanced Modes: General.
- Full Spectrum Access & Technical Experimentation: Amateur Extra.

All Amateur Radio operators in the U.S. must be licensed, with licenses valid for ten years before renewal. While the FCC regulates these licenses, the exams are administered by volunteer groups of Amateur Radio operators. Fun Fact: in 2007, the FCC stopped requiring Morse code proficiency as part of the licensing process.

I recommend starting with your licensing on the National Association of Amateur Radio's website. They will walk you through registering with the FCC and finding an exam location and date.

https://www.arrl.org/find-an-amateur-radio-license-exam-session

NAVIGATING POWER RESTRICTIONS

Navigating the power limits of your Baofeng radio is crucial for legal and efficient operation. Understanding these limits, set by the Federal Communications Commission (FCC), helps avoid interference with other services and prevents potential penalties. Here's a guide to help you manage power limits responsibly:

Understanding Power Limits: Power limits, expressed in watts (e.g., 1W, 5W, 50W), dictate the maximum transmitting power allowed on specific radio frequency bands. The FCC establishes these limits based on band propagation characteristics and the need to minimize interference between radio services (*Ham Radio Licenses*, n.d.).

Adjusting Transmit Power on Baofeng Radios: To change your Baofeng radios transmit power, typically on models like the UV-5R, follow these steps:

1. Press "MENU" to access settings.
2. Navigate to the "TXP" (Transmit Power) sub-menu using the arrow keys.
3. Enter the "Power" settings by pressing "MENU" again.
4. Choose between "High" (4-5 watts), "Mid" (2-2.5 watts, model dependent), or "Low" (1-0.5 watts) power settings.
5. Confirm and save your choice by pressing "MENU."

Consequences of Exceeding Power Limits: Exceeding FCC power limits is considered harmful interference and can lead to (*Enforcement Overview*, 2020):

- Substantial fines.
- Equipment confiscation.
- License revocation.
- In severe cases, imprisonment.

Best Practices for Power Management:

- Use the minimum power necessary for effective communication.
- On unlicensed bands like FRS/MURS, adhere to the 2W limit.
- For repeater use, higher power like 8W may be necessary.
- For short distances, 1-4W is sufficient and conserves battery life.

Typical Power Limit Examples:

- FRS: 2 watts maximum ERP.
- MURS: 2 watts is the maximum conducted power.
- GMRS: Up to 50 watts ERP, often limited to 5 watts on specific channels.
- Amateur Radio: Up to 1500 watts ERP on some HF bands for Extra Class licensees.

The Importance of Understanding Power Limits: Baofeng radios, capable of transmitting at higher power levels, require careful management to comply with FCC regulations. For instance, the popular UV-5R model can broadcast up to 5 watts, which exceeds the limits for unlicensed bands like FRS and MURS. Always verify FCC regulations for the intended frequency band and adjust your radio's power settings accordingly. This approach ensures compliance and helps you operate your radio within the legal framework, maintaining orderly and efficient spectrum usage.

LEGALITY BEFORE OPERATING

Before using your Baofeng radio, ensure you're up to date with FCC regulations, as these can change. Quick checks safeguard against illegal transmissions and promote responsible use. Remember, staying compliant protects you and respects the wider radio community.

The amateur radio community is an excellent resource for guidance, especially when in doubt about regulations. This collaborative learning fosters a compliant and interference-free radio environment.

Practicing diligence in adhering to rules helps you enjoy your Baofeng radio hobby without fear of regulatory issues. Be cautious and stay informed to avoid infractions.

———

As you master your Baofeng radio's basics, our next chapter will introduce advanced features, shifting you from beginner to proficient. Get ready to fully engage with your radio and delve deeper into the exciting realm of amateur radio, an invaluable skill for any prepper looking to enhance communication readiness.

The upcoming chapter will elevate your skills, transitioning you from a novice to a proficient Baofeng user. Prepare to unlock your radio's full potential.

7

ADVANCED OPERATIONS – BEYOND THE BASICS

F or the adept prepper, mastering advanced features of the
Baofeng radio is not just about enhancing technical skills; it's
about ensuring readiness and versatility in unpredictable situations.
This chapter takes you beyond the basics, equipping you with the
knowledge to utilize your Baofeng radio's full potential in scenarios
where efficient communication can be a game-changer.

You'll explore the nuances of dual watch for simultaneous channel
monitoring, understand the criticality of emergency transmission in
crisis situations, and learn effective memory channel management for
swift access to vital frequencies. Additionally, insights into battery
management and advanced troubleshooting will prepare you to maintain reliable communication under various conditions. These skills
are indispensable for any prepper looking to stay one step ahead in
emergency preparedness and effective response.

KEEPING AN EAR ON TWO CHANNELS: EXPLORING DUAL WATCH

Imagine being at an airport, listening in on the ground crew's chatter and the pilot's communication, or in the woods and listening as your two teams and basecamp communicate.

The Dual Watch/Dual Reception feature of your Baofeng radio offers a similar functionality. It lets you monitor two frequencies or channels, alternating between them, so you don't miss any important communication. This advanced capability is handy when staying informed on multiple channels is critical during an emergency.

How Dual Watch Works on a Baofeng Radio:

- First, set your radio to Frequency Mode (VFO/MR button)
- Next, input the first frequency or channel you want to monitor.
- Press the A/B key to switch to the other display line.
- Input the second frequency or channel.
- Press the "MENU" button on your Baofeng UV-5R.
- Use the arrow keys to navigate to the "TDR" (Dual Watch/Dual Reception) setting or directly press "7".
- Press "MENU" again to enter the TDR setting.
- Use the arrow keys to select "ON" for Dual Watch.
- Press "MENU" to activate Dual Watch.
- Press "EXIT" to return to the main screen.
- You'll see both frequencies displayed, and an 'S' will confirm Dual Watch is active. You are now monitoring multiple frequencies.

Understanding the Display:

- A steady arrow or pyramid symbol indicates the frequency you're set to transmit.

- A flashing arrow means you're receiving a transmission on that frequency.

Transmitting on Dual Watch:

- While Dual Watch allows you to receive on two frequencies, you can only transmit on one at a time (GMRS, n.d.).
- To switch the transmitting frequency, press the "A/B" button.

Practical Applications of Dual Watch:

- Emergency Monitoring: Keep an emergency channel on standby while using another for regular communication.
- Repeater Input Monitoring: Listen to a repeater's input and output frequencies.
- Team Coordination: Monitor both command and field team frequencies simultaneously.
- Enhanced Situational Awareness: Stay informed on two vital channels in dynamic environments.

Limitations and Considerations:

- Single Transmission: You cannot transmit on both frequencies simultaneously.
- Model Support: Not all Baofeng models support genuine Dual Watch. Verify your model's capabilities.
- Battery Life: Monitoring two channels can drain the battery faster.
- Distraction Potential: Continuous dual monitoring might be distracting, so use it judiciously.
- Practice Required: Effectively using Dual Watch takes practice to master rapid channel switching and timely transmissions.

In summary, Dual Watch is a versatile tool that, when used effectively, can significantly enhance your communication capabilities and situational awareness. Whether for emergency preparedness, team coordination, or just staying in the loop on multiple fronts, mastering Dual Watch can elevate your experience with your Baofeng radio. Remember to practice and familiarize yourself with this feature to maximize its benefits.

KEEP YOUR HANDS FREE BY UNLOCKING VOX

"Voice-Operated Transmission," or VOX, is a dynamic feature on many Baofeng radios that brings the convenience of operating hands-free. When VOX is enabled, your radio can transmit automatically in response to your voice, eliminating the need to press the Push-to-Talk (PTT) button for each communication. This feature is paramount when your hands are occupied, or a quick response is critical.

Activating VOX on Your Baofeng Radio: To enable VOX on a Baofeng radio, such as the UV-5R:

- Press the "MENU" button to access the radio's settings.
- Navigate to the "VOX" option (press '4').
- Enter the VOX settings ranging from Off, 1 (Low) to 10 (High) sensitivity.
- Start with Medium sensitivity (5) and adjust as needed to suit your environment.

Setting the Right Sensitivity Level: Getting the VOX sensitivity setting right is crucial:

- Too high, and the radio might transmit unintended background noise.
- If it is too low, your voice may not trigger the transmission effectively. Test the settings in your usual environment to find the sweet spot where your voice activates transmission reliably without false triggers.

Enhancing VOX with External Microphones: For improved VOX functionality, consider using an external headset, lapel, or throat microphone. These accessories are designed for clearer voice capture, reducing the chance of unintended transmissions caused by ambient noise.

The Benefits of Using VOX:

- Convenience: Frees up your hands for other tasks, as the radio transmits automatically when you speak.
- Natural Conversations: Provides a more fluid communication experience, removing the need for button pressing.
- Quick Response: Enables faster communication since it transmits as soon as you start speaking.
- Multitasking: Ideal for scenarios where you must communicate while your hands are busy.
- Accessibility: Beneficial for individuals who find it challenging to use PTT buttons.

Practical Scenarios for VOX Use:

- Climbing: Maintain communication without having to let go of your gear.
- Medical or First-Aid Situations: Communicate with others while attending to injuries or administering first aid, where your hands are occupied with medical supplies or patient care.
- Hiking or Driving: Keep your hands free for safety while staying connected.
- Wilderness Survival: Communication is essential when building shelters, starting fires, or preparing food, as both hands are needed for these survival tasks.
- Search and Rescue Operations: Stay in constant contact during search missions in challenging terrains, where hands-free communication is essential for coordinating efforts and ensuring safety.

- Fishing or Hunting: Maintain communication with partners while handling fishing rods, setting up baits, or aiming during a hunt.

Using VOX Responsibly: To maximize the benefits of VOX while minimizing drawbacks:

- Utilize a headset or lapel mic for focused voice pickup.
- Test and set the appropriate sensitivity level.
- Be aware that your radio is a "hot mic" when VOX is active.
- Disable extra tones like Roger Beep (a beep you hear after you release the PTT button) to prevent feedback loops.
- Monitor battery usage, as VOX can increase power consumption.

Mastering the VOX feature on your Baofeng radio can significantly enhance your communication experience. It offers a hands-free, responsive, and convenient way to stay connected. With some practice and the correct settings, VOX can transform how you use your radio, making it an invaluable tool in various scenarios.

RAISING THE ALARM WITH YOUR RADIO: EMERGENCY ALERT

The emergency alert feature on Baofeng radios is a potentially life-saving tool designed for critical situations. When activated, it signals a state of distress, alerting nearby radio users who can offer assistance. This feature is powerful but must be used responsibly and exclusively in emergencies.

Activating the Emergency Alarm: On most Baofeng models, such as the UV-5R, activating the emergency alarm is straightforward:

- Press the "CALL" button for about three seconds.

- This action triggers a loud, continuous beeping alarm tone and rapid flashing of the LED indicator, making the distress signal hard to miss.
- The alarm typically sounds for 30 to 60 seconds. To deactivate it manually before it auto-stops, press and hold the "CALL" button again for about two seconds.

Transmitting a Distress Signal: In addition to emitting an audible alarm, this feature also transmits a unique distress signal over the currently tuned channel. Other radios within range can pick up this transmission if tuned to the same frequency. Most Baofeng radios override volume or mute settings to ensure the distress call is heard (*Baofeng - Baofeng Troubleshooting*, 2022).

Practical Applications of the Emergency Alert: This emergency alarm can be crucial in various real-life scenarios:

- Hiking Accidents: Injured hikers can alert their group or nearby rescuers.
- Structure Fires: Notifying occupants of immediate evacuation needs.
- Medical Emergencies: Quickly signaling for help during health crises.
- SOS Signaling: Attracting attention when stranded in remote areas.

Responsible Use of the Emergency Alarm:

- Avoid False Alarms: Be cautious not to activate the alarm accidentally.
- No Pranks: Misusing the alarm for non-emergency purposes is unethical and potentially illegal.
- Confirm Alarms: Verify the legitimacy of any received emergency alarms before responding.

- Legal Compliance: Be aware of local laws regarding using emergency signals and adhere strictly to them.

When used correctly, the emergency alarm on your Baofeng radio is a critical feature that can significantly enhance safety in dangerous situations. Understanding and respecting its function is crucial, ensuring it's used appropriately and only in true emergencies. Doing so contributes to maintaining the alarm's integrity and effectiveness as a vital distress signal.

PERSONALIZING YOUR RADIO: MASTERING MEMORY CHANNELS

Memory channels on Baofeng radios are essential for personalizing and streamlining your radio experience. By saving specific frequencies into these channels, you gain quick, one-touch access to your most frequently used frequencies, bypassing the need for manual entry each time. This functionality tailors your radio to your unique needs, making its operation far more efficient.

Manually Programming Memory Channels on a Baofeng UV-5R: To manually save a frequency into a memory channel. Go back to Chapter 4 for detailed instructions.

Software Programming for Convenience: While manual programming is effective, software like CHIRP is a faster, more user-friendly way to program multiple channels. This method, detailed earlier in Chapter 4, allows bulk frequency input.

Benefits of Using Memory Channels:

- Quick Frequency Access: Easily switch to saved channels faster than manual input.
- Organized Frequencies: Group and organize channels logically, such as sequential numbering for local repeaters.

- Efficient Scanning: Scan through programmed channels to detect active transmissions.
- Customizable Setup: Tailor the channel lineup to your specific needs.

Tips for Optimizing Memory Channel Use:

- Logical Grouping: Organize channels in a way that makes sense for your usage, like grouping all local repeaters together.
- Naming Channels: Name channels for easy identification if your programming software allows.
- Separate Transmit/Receive Storage: Use pairs of memory channels to store separate transmit and receive frequencies for repeaters.
- Leave Space for Expansion: Skip some numbers between channel groups for future additions.
- Routine Scanning: Regularly scan through channels to stay informed of active transmissions.
- Back-Up Your Setup: Save your configuration externally for easy recovery.

Understanding Memory Channel Limitations:

- Limited Number: Baofeng radios typically offer 128–250 memory channels, which may require thoughtful management.
- Temporary Storage: Memory channels hold settings until overwritten or reset, necessitating periodic backups.
- Manual Selection Speed: Selecting channels manually can be slower than automatic scanning.
- Organizational Discipline: Effective grouping and naming are essential for quick and easy channel identification.

In conclusion, mastering memory channels is vital to becoming proficient with your Baofeng radio. These channels not only provide a fast

way to access your preferred frequencies but also allow you to organize and customize your radio for optimized use. With practice and
thoughtful setup, you'll find memory channels indispensable in your
everyday radio operations.

YOUR RADIO'S LIFELINE: BATTERY LIFE

Batteries are the lifeline of your Baofeng radio, providing the power
needed for all its functions. Understanding how to manage and maintain your radio's batteries effectively ensures long-lasting performance and reliability.

Types of Batteries for Baofeng Radios: Most Baofeng radios use
rechargeable lithium-ion (Li-ion) batteries. These batteries are known
for their high energy density and long life span. They are typically
lightweight, making them ideal for handheld radios like the Baofeng
models.

Maximizing Battery Performance:

- Initial Charge: When you first get your Baofeng radio, charge
  the battery fully before using it. This ensures optimal battery
  performance.
- Regular Charging: Li-ion batteries do not suffer from
  memory effects, so it's best to charge them regularly and keep
  them topped up.
- Deep Discharge: Avoid completely draining the battery, which
  can shorten its life. Recharge the battery when you notice a
  decrease in performance.

Understanding Battery Capacity:

- Baofeng batteries are often rated in milliampere-hours (mAh),
  indicating how much charge they can hold. Higher mAh
  ratings mean longer battery life per charge.

- For instance, an 1800 mAh battery will last longer than a 1500 mAh battery under the same usage conditions.

Charging Your Baofeng Battery:

- Use the charger provided with your radio or a compatible replacement.
- Avoid overcharging the battery; remove it from the charger once fully charged.
- Some Baofeng models come with charging cradles for convenient charging.

Storage and Care:

- If you're not using your radio for an extended period, store the battery separately in a cool, dry place.
- Avoid exposing the battery to extreme temperatures, as heat can degrade its performance and lifespan.

Replacing the Battery:

- Over time, you may notice a reduction in battery life. This is normal for rechargeable batteries and indicates it's time for a replacement.
- Use only Baofeng-approved or reputable third-party batteries to ensure compatibility and safety.

In conclusion, properly managing your Baofeng radio's battery ensures that it remains a reliable communication tool. Regular charging, appropriate storage, and timely replacement are vital for maintaining battery health. Understanding and caring for your radio's battery allows you to enjoy extended use and avoid interruptions in communication.

ADVANCED TROUBLESHOOTING

As you delve deeper into your Baofeng radio's use and explore its advanced features, you might encounter complex issues that call for more sophisticated troubleshooting techniques. Here's a guide to help you navigate some common advanced problems and their potential solutions:

Troubleshooting Tone Squelch Activation Failures: If you're using CTCSS or DCS tone squelch settings and suddenly find yourself unable to receive signals, the issue often lies in mismatched codes between your radio and the transmitting radio. For successful communication:

- Ensure both radios use the exact same CTCSS subaudible tone frequency or DCS code values.
- Double-check the squelch settings to confirm that your radio's tone frequency or code value precisely matches the transmitting radio's settings.

Addressing Repeater Connection Issues: If you're experiencing difficulties connecting to repeaters or if transmissions are broken or distorted, consider these checks:

- Confirm that the transmit offset in your radio matches the repeater's input frequency.
- Check that any required tone encode (usually CTCSS) for the repeater is correctly set on your radio.

Managing Shorter Battery Life: A rapidly depleting battery can be a sign of various issues:

- Avoid exposing the radio to extreme temperatures.
- Regularly inspect the battery for swelling, which indicates damage.

- Ensure consistent and complete charging after use.
- Use original Baofeng batteries and chargers for optimal compatibility and performance.

Solving Bluetooth Connectivity Issues: For problems pairing with Bluetooth devices, these steps can help:

- Engage pairing mode on your radio and the Bluetooth device while ensuring they are close.
- Minimize physical obstructions between the radio and the Bluetooth device.
- Update your radio with the latest firmware to enhance Bluetooth stability.
- Consider resetting both the radio and the Bluetooth accessory to factory settings before attempting to pair again.

Leveraging Community Expertise: Don't hesitate to seek help from the Baofeng user community for advanced troubleshooting:

- Online Communities: Platforms like Reddit's Baofeng subreddit and Facebook groups like "Baofeng Radio Club" are valuable places to share troubleshooting advice.
- Manufacturer Forums: Official forums like BaofengTech can provide specific guidance for advanced issues.
- Local Hamfest Meets: Meeting experienced users in person can offer direct, hands-on assistance.
- YouTube Tutorials: Visual guides from experienced users can be beneficial for complex troubleshooting.

Remember, while learning to solve advanced issues independently is rewarding, tapping into the collective wisdom of the Baofeng community can save you time and frustration. Over time, this shared knowledge will significantly enhance your understanding and proficiency in handling your Baofeng radio.

EMPOWERED THROUGH CONTINUOUS LEARNING

Mastering your Baofeng radio is an ongoing journey of learning and adaptation. As technology evolves with new models, enhanced features, and expanded capabilities, staying informed and engaged with the Baofeng community is key to honing your skills.

Stay Informed Through Official Sources:

- Regularly visit Baofeng manufacturers' websites, such as BaofengTech and BaoFengRadio, to stay up-to-date on the latest products, updates, and accessories.
- Subscribe to their newsletters and follow their social media channels for the latest news.

Enhance Your Operator Skills:

- Consider obtaining an amateur radio license. Earning your license and call sign is a badge of commitment and expertise.
- Even if you choose not to get licensed, continuously improve your operator skills. Practice navigating your radio's menu and settings daily. Familiarize yourself with programming memory channels and experiment with different accessories to optimize your setup.

Connect with Local Clubs:

- Local amateur radio clubs are invaluable resources. Regular meetings offer a chance to learn from experienced hams. These clubs can provide practical advice, legal best practices, equipment recommendations, and hands-on training.
- Participating in club activities and events also helps you build a network of fellow radio enthusiasts.
- In these groups, you will meet and become friends and build networks that will be valuable assets during a crisis.

Stay Proactively Engaged:

- Actively participate in online forums and platforms dedicated to Baofeng radios. Engaging with fellow users allows you to tap into a wealth of collective experience and knowledge.
- Share your own experiences and learnings. The more you contribute, the more you gain from the community.

After mastering the advanced aspects of your Baofeng radio, it's time to explore one of its most vital uses: emergency and disaster communications. When traditional communication infrastructures fail, your Baofeng radio becomes a crucial link to the outside world. The next chapter will delve into effectively using your radio in times of crisis, ensuring you're prepared to communicate when it matters most.

8

STAYING CONNECTED – BAOFENG RADIOS DURING DISASTERS

In times of disaster, when traditional communication networks fail, Baofeng radios emerge as crucial tools for maintaining vital connections. This chapter delves into the indispensable role of Baofeng radios during emergencies, offering guidance on preparation, effective communication, and maintenance.

Baofeng radios stand out in emergencies due to their resilience. These battery-powered devices enable direct radio wave transmissions over considerable distances, providing a reliable alternative when standard communication infrastructures collapse. Their ease of network setup allows for quick and efficient coordination during crises.

Understanding the tactical advantage of Baofeng radios is essential. We will explore their functionality, adaptability, and decentralized operation, which are crucial for effective crisis management. This chapter will equip you with practical knowledge for using Baofeng radios during emergencies – preparing your device, maintaining it, and best practices for communication.

We aim to empower you with the skills to use Baofeng radios confidently in emergencies. You'll gain insights into transforming these devices from simple communication tools into strategic assets for emergency response. Whether you're a first responder or a concerned citizen, this knowledge is vital for ensuring preparedness and effective response in times of need.

NATURAL DISASTERS

Mother Nature is beautiful yet unpredictable. A sunny day can quickly turn into a stormy night, a calm sea into a raging torrent. Natural disasters such as hurricanes, earthquakes, floods, and wildfires can strike without warning, disrupting normal communication channels and isolating communities. In such situations, your Baofeng radio can serve as a critical lifeline.

With its ability to operate independently of mobile networks and internet connections, your Baofeng radio can transmit and receive crucial information when conventional communication methods fail. Whether receiving updates on the disaster situation, requesting assistance, or coordinating relief efforts, your Baofeng radio can help you stay connected and informed even in the heart of a natural disaster.

Hurricane Katrina

Picture the scene: a city submerged, homes wrecked, and communication networks crippled. This was the grim reality of New Orleans in 2005, in the aftermath of Hurricane Katrina. Amid the chaos and destruction, Baofeng radios emerged as an unexpected hero. As cell towers and internet lines succumbed to the storm, these sturdy radios kept their signal, providing a lifeline to stranded residents and first responders alike.

Emergency teams used their Baofeng radios to coordinate rescue operations, communicate with relief centers, and provide real-time updates on the situation. Citizens used them to request help, share information, and stay connected with their loved ones. In a city silenced by the storm, the airwaves buzzed with the static and chatter of Baofeng radios, echoing the resilience of the human spirit.

California Wildfires

Now, let's shift our gaze to the golden state of California, known for its sunny beaches, Hollywood glamour, and, sadly, devastating wildfires. In 2018, the state faced one of its deadliest wildfire seasons. As the flames spread, engulfing homes and forests alike, Baofeng radios again stepped into the fray.

Firefighters used their Baofeng radios to communicate with each other, coordinate their efforts, and navigate the dangerous terrain. Evacuees used them to receive updates on the fire's progress, find safe evacuation routes, and stay connected with the outside world. Amid the smoke and flames, the humble Baofeng radio emerged as a beacon of hope, its signal cutting through the haze of uncertainty.

Wildfires can be extremely deadly and fast. They can overtake an area with only a minute warning based on wind patterns. This is when you need to activate your sudden evacuation plan. A bug-out bag is crucial for disaster preparedness, and alongside your Baofeng radio, it should include essentials for at least 72 hours of survival. Ensure it contains water purification methods, non-perishable foods, clothes, a basic first-aid kit, and tools like a multi-tool and flashlight. Include important personal documents, cash, and maps for navigation. Tailor it to your needs, check it regularly, and keep it ready for immediate departure.

SEARCH AND RESCUE OPERATIONS

Picture a game of hide-and-seek in a vast forest. The seeker calls out, and the hider responds, their voices guiding each other through the dense undergrowth. In search and rescue operations, your Baofeng radio plays a similar role. It's a voice in the wilderness, a beacon in the dark, and a critical tool in locating and rescuing lost or injured individuals.

Search and rescue teams often operate in remote and challenging terrains where regular communication systems may not be reliable. Baofeng radios, with their robust performance, long battery life, and ability to communicate over large distances, can provide a reliable means of communication. Coordinating search efforts, sharing information about the missing person, or calling for medical help every transmission can be a step towards a successful rescue.

Configure a "Grab and Go" Kit

Assemble a self-contained Go Bag for your handheld with essential accessories for emergency use. Include spare charged batteries, a 12V cigarette lighter charger, a foldable external antenna, a printed reference sheet of programmed channels, headsets, and a small notepad and pen for written messages. Keep this kit prepared and easily accessible for rapid deployment when needed. Your radio and all its essentials.

Nepal Earthquake

Lastly, we journey to the foothills of the Himalayas, to Nepal, a country known for its stunning landscapes and rich culture. In 2015, a massive earthquake struck Nepal, causing widespread destruction. As the ground shook and buildings crumbled, communication networks collapsed, leaving millions cut off from the rest of the world.

Once again, Baofeng radios rose to the occasion. Rescue teams, relief workers, and even ordinary citizens turned to these radios as their primary means of communication. They used them to coordinate relief efforts, call for medical assistance, and provide comfort and reassurance to those affected by the disaster. In the face of unimaginable devastation, Baofeng radios served as a lifeline, connecting hearts, hands, and hopes.

Earthquakes destroy city infrastructures. Roads, bridges, and ways out of the impacted area are blocked. Supply chains and emergency services may be unavailable for days. Electricity might be out for weeks. Your stocked home, base station, and Baofeng will be invaluable here.

POWERING YOUR RADIO WITH A CAR INVERTER

When the power grid fails, as it most likely will after an earthquake, your car can become a valuable asset for transportation and a power source for your Baofeng radio and other small electronics. Here's how to turn your vehicle into a makeshift power plant with the help of an inverter.

Utilizing a Car Inverter

For Light Loads: A cigarette lighter socket in your car typically supports up to 150 watts, sufficient for charging phones and radios, powering small LED lights, or running a compact fan.

For Heavier Loads: If you need to power devices that require more energy, you'll need to clamp an inverter directly to the car battery. Mounting the inverter on a block of wood is advisable to prevent it from moving and ensure safety.

- Remember that your car's alternator generates about 400-600 watts. Therefore, if you operate appliances requiring more power than this, they will drain the battery after the alternator's output is exceeded.
- Ensure you are idling your vehicle during these types of loads.

Efficient Use: Your car, equipped with an inverter, is ideal for powering household items within the lower power range. It's impractical to start a generator to charge your phone or radio; use your car's battery instead.

Fuel Efficiency: A car or truck with a full gas tank can idle and provide power for approximately 24 hours, making it an efficient emergency power source.

Safe Fuel Storage for Extended Power Supply

Fuel Storage: Invest in several 5-gallon fuel cans for gasoline storage. This fuel can be used for your car or a backup generator.

- Monthly Rotation: Fill up one can each month and mark it with the month's number.
- Yearly Cycle: After storing the fuel for a year, rotate it into your vehicle's tank. This ensures the fuel doesn't age too much and keeps your stock fresh. Pour the can with this month's number into your car. Then go fill up your gas can with new gas.

Following these guidelines, you can ensure you're prepared for power outages with a reliable and efficient system for powering essential devices like your Baofeng.

POWER OUTAGES

Imagine a sudden blackout, an unexpected power outage. The lights go out, computers shut down, and mobile networks go offline, plunging the world into a communication blackout. In such scenarios, your Baofeng radio can be a beacon of light, enabling you to stay connected even when the power grid fails.

During power outages, Baofeng radios can operate on their battery power, allowing you to receive critical updates and communicate with others. Whether contacting family members or local authorities or coordinating community response efforts, your Baofeng radio can ensure you're never truly in the dark.

Power outages are unpredictable, and service restoration depends on what happened and whether the electric company has workers willing to leave their families to reconnect services. This is when your extra batteries, battery bank, car inverter, generator, and fully stocked fuel rotation come into play with your Baofeng.

A home battery bank can be invaluable in a power outage, especially for keeping critical communication lines open, such as charging your Baofeng radios. Here's how to create and utilize a battery bank tailored for small-scale energy needs.

CREATING A SIMPLE HOME BATTERY BANK

Begin by purchasing the following components:

- **Deep Cycle Marine Battery**: Choose a Group 31 or 29 Flooded Lead Acid Battery, with a capacity of around 75 Amp-hours, to ensure enough storage for your needs.
- **Intelligent 3-Stage Computerized Battery Charger**: This charger will optimize your battery's health and longevity, adjusting the charge as needed. Opt for a 10 AMP charger to balance speed and battery care.
- **Voltmeter**: Essential for monitoring your battery's charge level and overall health. The voltmeter needs to be the type that can be plugged into an outlet.
- **Inverter**: A simple 150-to-400-watt converter is more than enough for our purposes. The inverter, which will have clamps that connect directly to the battery, converts the stored DC power in your battery to AC power for your devices.
- **Compact Table Lamp with Low Wattage Bulb**: This low-energy lighting solution can run off your battery bank, so you won't need a flashlight when you go to your battery bank.

- **Orange T-shape 1 Plug to 3 Outlet Splitter Adapter:** Buy several. They allow you to turn your one outlet plug into three outlet plugs.

Setting Up Your Bank

- Mount the battery onto a table, shelf, or workbench in a dry, temperature-controlled location.
- Place the battery charger next to the battery and plug it into an AC outlet. Connect the charger to the battery.
- Place the inverter next to the battery and connect it directly to the battery. You now have power!
- Connect the voltmeter to the inverter. This will provide you with a visual display of the health of your batter.
- Connect your orange T-shape splitters so that you have more outlets to use.
- Connect your table lamp for light.
- In one of your orange splits, connect your Baofeng Radio charger.

Your bank is now set up and ready!

Using Your Battery Bank

With a 120-watt load, your battery bank should last 4-6 hours of continuous use. However, you don't need to use it continuously. That's not their purpose. The purpose is to charge your devices and disconnect from the bank once fully charged. Doing this will extend its functional use.

Remember, it's designed for small devices such as:

- **Yes**: Radio chargers, lights, phones, and tablets. To charge them to full health.

- **No**: Refrigerators, freezers, or coffee machines. Don't even think about running these devices.

Recharging Your Battery Bank

To recharge:

- Jump Start the Battery from Your Car: Jump the battery as usual. This method is suitable if you need a quick charge to keep essential devices running.
- From a Car Inverter: Connect your battery charger to an inverter attached to an idling car using an extension cord.
- From a Generator or Car Inverter: Connect your battery charger to a generator and let it recharge the battery.

Keep in mind that the battery bank is primarily for communications and lighting.

By following these steps, you'll have a reliable power source for your most critical devices, ensuring you can stay informed and connected even when the grid is down. Remember, the proper preparation can make all the difference in an emergency.

CIVIL UNREST

Communication plays a crucial role in times of civil unrest. It's a thread that holds communities together, a channel that disseminates critical information, and a platform that gives voice to the voiceless. Whether it's a peaceful protest, a political rally, or a sudden outbreak of violence, your Baofeng radio can be an indispensable communication tool.

Baofeng radios can help you stay updated on the situation, receive instructions from authorities, or coordinate with fellow community members. Operating independently of cellular networks and the internet ensures you can maintain communication even when these regular channels are disrupted.

Remember your time at your local ham club and setting up your neighborhood net? This is when your investment in time pays off.

OUTDOOR ADVENTURES

Venturing into the great outdoors is exhilarating but comes with risks. A simple hike can become a survival situation, a fun camping trip into an unexpected night in the wilderness. Your Baofeng radio can be valuable for your survival kit in such situations.

Whether lost on a trail, injured during a climb, or stranded due to bad weather, your Baofeng radio can help you call for help, receive weather updates, or stay connected with your group. It's not just a radio; it's your lifeline in the wild.

In the face of an emergency, your Baofeng radio is more than just a device; it's a lifeline, a beacon of hope. It's a testament to our resilience and capacity to adapt, innovate, and turn a simple communication tool into a lifeline. So, as you hold your Baofeng radio, remember: it's not just about the frequencies it can tune into or the signals it can transmit. It's about the lives it can touch, the hope it can inspire, and the difference it can make in times of need.

INDEPENDENT LOCAL RADIO NETS

Baofeng's flexibility, portability, and ease of use allow setting up independent local "radio nets" when needed (Cribbs, 2019a). A radio net is a network of stations communicating on a shared channel during an emergency.

Nets can quickly share urgent news and weather, safety alerts, request medical assistance, coordinate logistics like equipment and volunteers, and check on vulnerable community members. Having access to such information can be lifesaving.

Your chance to barter begins now! With this group of like-minded individuals.

PREPARING YOUR BAOFENG RADIO FOR EMERGENCIES

With some planning and preparation, your Baofeng radio can provide a resilient communication lifeline when standard infrastructure fails. Follow these tips to ensure your handheld is ready for disasters:

Program Emergency Frequencies

Use PC software like CHIRP to pre-program your region's radio with critical emergency frequencies. Good sources are NOAA weather radio, FEMA, Red Cross, local fire/police, and Amateur Radio emergency networks. Having these ready to select instantly saves precious time when a crisis strikes.

National Weather Service (NWS) Channels

The National Weather Service (NWS) operates a network of radio stations known as NOAA Weather Radio (NWR), which broadcast continuous weather information 24/7. These channels are real-time weather updates, keeping you informed about changing weather conditions, forecasts, and severe storm warnings.

Your Baofeng radio can tune into these NWS channels, ensuring you have access to this vital information. This can be particularly useful in emergencies, enabling you to stay ahead of the weather and make informed decisions.

The Emergency Alert System (EAS)

EAS plays a similar role in the modern world. It's a national public warning system that allows the President and local authorities to communicate emergency information to the public via television, radio, and other communication devices.

Your Baofeng radio can receive these EAS alerts when tuned to an appropriate frequency. This ensures you're promptly informed of local or national emergencies, enabling you to take appropriate action.

Install a Base Station

Baofeng radios, known for their versatility and affordability, can be an excellent choice for establishing a base station radio setup. Unlike their portable counterparts, base station radios are stationary units, typically placed on a desktop, and often include an integrated power supply within a frame alongside an external antenna. While you sacrifice portability, you gain significant advantages in power and range. Typically, these radios boast transmit power levels between 25 to 50 watts. Installing an outdoor antenna as high as possible is advisable to maximize your communication reach. However, if your range needs are modest, a simpler solution, like a mag-mount mobile antenna on a metal surface, such as a file cabinet, can suffice.

Installing a Baofeng base station at your home or bug-out location provides a steadfast communication hub, particularly critical if the power grid fails. You can effectively connect with aid and resources by operating on simplex or repeater frequencies covering 20–50 miles. Moreover, ensuring your base station is equipped with battery backup power is crucial for maintaining communication during power outages. By understanding these essentials, beginners can set up a reliable and efficient base station, leveraging Baofeng's capabilities for robust local communication.

Waterproof Your Handheld

Consider a waterproof carry case or plastic sealing bag for your Baofeng to protect it from water damage if caught out during flooding or storms. Ensure any accessories like batteries and microphones are likewise protected from moisture ingress.

The last chapter on maintenance includes what to do and how to care for your wet radio.

One of the benefits of Baofengs is their price. Being very affordable, you can, and should, often buy several and keep some in reserve as backups.

Hold Practice Drills

Schedule periodic emergency drills to rehearse your family's, neighborhood's, or team's disaster communication plan. Identify any equipment, training, or procedural knowledge gaps to address before an actual crisis. Consistent practice cements skills and readiness.

A question to ask yourself: Does your family know how to use your radio? Or are you the bottleneck and gatekeeper? If you get hurt, can anyone at your home use the radio to call for help or establish contact with others?

Learn and Grow

Expand your knowledge through local ham radio club workshops, online training programs, and FEMA emergency communication courses. Consider training for volunteer emergency roles like Community Emergency Response Teams (CERT), which provide crucial radio support.

- CERT (Community Emergency Response Team): CERT is a program that educates volunteers about disaster preparedness for hazards that may impact their area. It trains them in basic disaster response skills, such as fire safety, light search and rescue, team organization, and disaster medical operations. CERT aims to enable volunteers to assist others in their neighborhood or workplace following an event when professional responders are not immediately available to help.

The easiest way to find a local CERT team is to type 'Find Local CERT' into Google. Then click on the FEMA website link, which will bring you to the exact page you need.

Hardening Your Station

Professional-grade station surge suppressors protect your radio equipment against lightning strikes and power surges. Consider shielding from solar flares and EMP (Electromagnetic Pulse - a burst of electromagnetic radiation from certain natural occurrences like lightning or man-made events such as a nuclear explosion) with simple Faraday solutions like conductive mesh bags. Redundancy is key. Here are some more detailed steps on hardening your amateur radio station:

- Install whole-house surge suppressors or lightning arresters to protect all equipment. Look for units with high joule ratings.
- Use coax surge protectors rated for DC-1500 MHz on antenna feedlines (also known as transmission lines) to bleed static charges. These play a crucial role in connecting the radio transmitter and receiver to the antenna; feedlines carry radio frequency (RF) signals between the radio equipment and the antenna with minimal energy loss. Replace them periodically.

It's time for a quick deep dive on feedlines and bleeding static charges.

Static Charges on Antennas: Antennas, especially those outdoors, are prone to accumulating static electricity. This can be due to various environmental factors like wind, rain, or even the natural build-up of static in the atmosphere. Static charges can become problematic because they can create noise in the receiver, potentially damaging the radio's sensitive electronic components.

Bleeding Static Charges: To protect the radio equipment, it's crucial to have a method of safely discharging these static charges. This process is often referred to as "bleeding" static charges. It provides a path for the static electricity to discharge safely to the ground without passing through the radio's circuits.

Implementation: This can be achieved by grounding the antenna system. Grounding involves connecting the metallic parts of the antenna and feedline system to the earth, creating a direct path for static charges to flow into the ground. Proper grounding not only helps in bleeding off static electricity but also enhances the overall performance of the antenna and provides a safety mechanism against potential lightning strikes.

- Plug all station accessories like tuners, amplifiers, and computers into power strips with built-in surge protection.
- Unplug antennas and power when not in use to minimize the chances of damage—disconnect feedlines during storms.
- Store radios in conductive mesh Faraday bags when not in use to shield them from electromagnetic pulses.
- Build a low-cost Faraday cage from wire mesh and ground it appropriately to store backup gear.

It's time for a quick deep dive into Faraday cages.

- A Faraday cage is a protective enclosure made from conductive material, such as metal, which shields its contents from external electromagnetic fields. Named after Michael Faraday, who invented it in 1836, this cage works by redistributing electromagnetic charges around its exterior, effectively neutralizing electromagnetic radiation like static, radio waves, and EMPs (Electromagnetic Pulses) inside the cage. It is commonly constructed from a metal mesh, the size of which is smaller than the wavelength of the blocked radiation.

Other Tactics:

- Keep spare radios in Pelican cases (Pelican is a brand of rugged, durable cases designed to protect sensitive equipment from various kinds of damage, including impacts, dust, and water. These cases are widely known for their strength and reliability) lined with copper tape to create portable shielded boxes.
- Use surge-protected power strips and 12V DC power options for off-grid flexibility.
- Install gas tubes or spark gap arresters on antennas tuned to shortwave bands for added protection.

  - Gas tubes and spark gap arresters are surge protectors that safeguard electronic equipment from high-voltage surges. Gas tube arresters have a gas-filled tube that becomes conductive under high voltage, diverting the surge to ground, while spark gap arresters use two electrodes with an air gap, where a surge creates a spark to discharge excess voltage. Both types quickly respond to voltage spikes.

- Follow proper grounding procedures for all station equipment and consider multiple grounding rods.
- Check your homeowner's insurance policy regarding amateur radio equipment coverage. If a disaster hits and society doesn't collapse, you at least get your damaged equipment reimbursed by the insurance company.

Analog Backup

While convenient, strictly digital modes may be disrupted in disasters. Maintain analog capabilities like AM/FM transmission/reception for redundancy. You may be using analog radio without even realizing it. If you like to listen to the radio in your car, any FM station transmits audio to you via the analog system. Broadband antennas and older transceivers provide backup utility. Here are some more detailed steps for maintaining analog backup capabilities:

- Keep spare antenna parts like wire, insulators, and baluns (BALanced to UNbalanced: Baluns are devices that convert between balanced and unbalanced signals in radio; they are crucial for minimizing interference and matching impedance. They are used in connecting antennas to radios) to construct emergency broadband antennas. Ladder lines, long wires, and dipole antennas can cover large frequency ranges.
- Set aside basic analog equipment for backup, such as AM/FM portable radios and transceivers. Look for self-powered and weather-resistant models.
- Learn to tune and operate vintage analog transmitters like tube-based AM (Amplitude Modulation) and SSB (Single Sideband) equipment. These can transmit voice with just power and a wire antenna.

AM and SSB are two different methods of modulating radio signals for voice communication in ham radio. AM is simpler but less efficient, while SSB, by transmitting only a single sideband, offers greater efficiency and is better suited for long-range communication.

- Stock spare tubes, crystals, dynamotors, and other analog parts that may be hard to source later. Learn how to service and repair old gear.

- Practice using CW (Continuous Wave) and Morse code skills on HF bands. In the worst case, CW may work when voice modes fail.

CW is a mode of communication that uses a continuous wave signal to transmit information, typically in the form of Morse code.

Morse code is a system of encoding textual characters as standardized sequences of two different signal durations, known as dots and dashes.

In summary, CW is the method or mode used in radio communication, often employing Morse code as the language for transmitting messages.

- Establish an emergency cache of batteries, generators, fuel, and solar power options to keep analog stations transmitting if the grid goes down.
- Consider getting licensed for shortwave bands to transmit international AM/SSB signals during crises.
- Drill with local radio groups to practice passing voice and Morse code messages using vintage analog equipment. Build skills and experience.
- Become an Elmer, an affectionate term for an experienced amateur radio operator who mentors, guides, or assists newcomers to the hobby. Document your expertise to act as a resource for others.

Maintain Your Gear

Check emergency kit supplies for expiration or deterioration twice annually. Inspect antennas and cables for damage. Cycle batteries monthly. Keep all equipment ready for immediate use when required.

That was a lot! Don't worry about completing all of it right now. Focus on the basics and the easy things first. As you learn and gain experience, you can advance to the higher-level preparations.

ESTABLISHING EFFECTIVE COMMUNICATION DURING EMERGENCIES

When normal communication channels fail in a crisis, your Baofeng radio becomes invaluable for reliable updates and coordination. However, ineffective use can hinder response, so follow emergency communication best practices.

Listen More Than Transmit

During disasters, designated radio frequencies carry critical traffic from responders and agencies. Keep transmissions brief and focused only on vital information. Avoid tying up the channel with casual chatter. Listen attentively to stay continuously updated on the evolving situation.

Remain Calm and Clear

Speak slowly, calmly, and clearly. Take your time with your transmission despite the urgency. Identify yourself and your location or situation. Repeat critical information to confirm receipt. In high-stress situations, a measured delivery ensures your message is understood.

Conserve Battery Power

In prolonged emergencies, conserve handheld radio battery power through judicious transmit times, minimal standby monitoring, and low audio volume. Have spare charged battery packs and emergency chargers readily available. Manually turn off the radio to avoid draining batteries in standby mode. But periodically check emergency channels for critical updates.

Avoid Overusing Alert Tones

The emergency alert tone on Baofeng radios should only be used for immediate, life-threatening crises requiring urgent response. Do not use this alert for non-critical issues, leading to alert fatigue. Always cancel the alert once the escalation call is acknowledged.

MAINTAINING YOUR BAOFENG RADIO IN SEVERE CONDITIONS

Emergencies often entail operating radios in harsh environments very different from everyday conditions. Situations like storms, floods, fires, or extreme cold present unique challenges for maintaining and protecting your equipment. Let's explore valuable tips for keeping Baofeng handhelds functioning in demanding conditions.

Water and Moisture Resistance

Floods, storms, and heavy rainfall can expose radios to soakings that risk internal corrosion and electrical shorts. There are several protective measures you can take:

- Use a waterproof case or dry bag to shield the radio, securing it well with o-ring gaskets. Some purpose-built radio cases even allow the operation of the controls through a transparent membrane (*Baofeng UV-5R+ Plus Review*, 2022).
- For lightweight protection, seal the radio in a plastic bag or food wrap, securing the open end with a rubber band. However, this won't permit access to the controls (*Baofeng UV-82HP vs UV-5R*, 2022).
- Apply corrosion-inhibiting gel or grease on any exposed metal plugs and contacts if water intrusion is expected. Avoid greasing plastic parts (*Storage of a Baofeng UV-5R*, n.d.).

- Use a speaker microphone accessory to keep the radio in a bag or case while allowing push-to-talk operation. Just ensure the accessory jack is well sealed (Baofener, 2023a).
- If your radio does get wet, remove the battery immediately and allow the unit to dry thoroughly for at least 24 hours before reconnecting the power (Baofener, 2023b).

Dust and Dirt Resistance

Mud, dust, and sand storms can wreak havoc on radio joints, buttons, and exposed jacks. To protect from gritty grime:

- Start with a waterproof case, sealed plastic bag, or quality holster as your first line of defense (Twowayradiogear, 2014).
- Periodically blow out any particles from crevices using short bursts of compressed air (*How to Preserve Your Two Way Radios*, n.d.).
- Use rubbing alcohol and cotton swabs to remove dirt from tight spots, avoiding harsh scrubbing (*How to Clean and Care for Your Two-Way Radio*, n.d.).
- After severe grime exposure, disassemble the radio following the manufacturer's steps and thoroughly clean all components (Twowayradiogear, 2014).

Temperature Extremes

Both hot and cold temperature extremes can impact handheld radio function. To mitigate the damage:

- In hot weather, store the radio in the shade and use passive cooling methods, such as placing it near cool running water. Actively cooling an overheated radio can cause further problems from condensation.

- In cold, keep the radio insulated as much as possible and ideally close to your body heat. Avoid charging batteries in extreme cold.
- Whether hot or cold, keep the radio off when not in actual use to prevent compounding any temperature damage. The unit will assume ambient conditions.
- In vehicles, avoid direct sunlight exposure on the dash and use shade, tinting, or reflective barriers to prevent excessive heat buildup.
- Select lithium-ion replacement batteries when available, as they handle temperature fluctuations better than NiMH batteries.

Care During Power Outages

When normal electrical service is disrupted, take precautions when charging handheld radios to avoid voltage spikes or surges when power is restored:

- Avoid using uninterruptible power supplies (UPS) or standalone generators for charging. Voltage regulation may be poor.
- If you are charging the radio using a vehicle, disconnect the charger when the engine starts to avoid peak alternator output.
- When power is restored, wait 5–10 minutes before reconnecting any charger, allowing grid voltage to stabilize.
- Select power strips or chargers with built-in surge suppression specifically rated for electronics.

Handling Corrosion Issues

Remove the batteries immediately if your radio is exposed to moisture during use, and dry the radio thoroughly with a microfiber cloth. Do not use hair dryers or other heat sources, as they can damage the internals. If humidity is an issue, place moisture-absorbing packs in the storage box.

Check for any signs of corrosion like discoloration, pitting, or deposits around metal components and solder joints. Corrosion inhibitor gels and contact sprays can protect susceptible components. If corrosion is severe, replace affected parts.

Storage Best Practices

- Store handheld radios in impact-resistant, waterproof cases with padding to protect them from physical damage. Avoid fabric bags that allow moisture ingress.
- Remove batteries from the radio and charger before storage to prevent battery drain or acid leaks.
- Place several silica gel desiccant packs in the storage container to actively absorb moisture from the air and prevent humidity damage.
- Avoid storing radios in non-climate-controlled environments with extreme temperature and humidity fluctuations like attics or sheds.

Anticipating the harsh conditions likely present during emergencies allows you to properly prepare your Baofeng radio to withstand the elements. Protecting the unit from moisture, grit, and temperature extremes keeps it functioning when needed. Follow preventative maintenance steps; your hearty handheld will provide reliable communication through any challenges ahead.

From the submerged streets of New Orleans to the fire-ravaged land-scapes of California to the earthquake-shattered towns of Nepal, Baofeng radios have repeatedly proven their mettle. They've shown that in times of crisis, when all else fails, communication endures. And as you hold your Baofeng radio, know that you own more than just a device. You have a lifeline, a beacon, a voice in the wilderness, ready to spring into action when duty calls.

Now, let's continue to delve deeper into the world of Baofeng radios. In the upcoming chapter, we'll explore how you can use your Baofeng radio in your everyday life. You'll discover how your Baofeng radio can enhance daily communication, from neighborhood watch programs to local ham radio clubs. So, stay tuned, and let's continue this together.

'The Command Packet' includes a Morse code cheat sheet and the 3-3-3 Radio Plan, which can be activated during SHTF events.

www.MorseCodePublishing.com/Command

9

BAOFENG UNLEASHED - EVERYDAY REAL-WORLD COMMUNICATION

The actual test of your Baofeng radio mastery is more than just understanding its functions; it's about applying those skills in real-world scenarios where effective communication can be a life-saver. This chapter focuses on practical applications of your Baofeng radio, especially in situations that preppers often encounter.

You'll learn how to use your radio for outdoor activities like hiking and camping, where reliable communication can keep you safe and connected. The chapter also covers using Baofeng radios for event coordination and off-grid communication, essential skills for managing community events, or navigating through areas without cellular coverage. These real-life applications of Baofeng radios enhance your outdoor experiences and prepare you for handling emergencies with confidence and expertise.

COMMUNICATING IN THE GREAT OUTDOORS

Baofeng radios are ideal for communicating during outdoor activities where cell phone coverage is unreliable or nonexistent. Their simplicity, durability, and flexible power options make them well-suited for treks into the wilderness. Let's explore typical usage scenarios where a Baofeng can keep you connected.

Grab 'The Command Packet,' which includes 10 radio operational checklists for real-life scenarios to ensure your radio is always ready and functional.

www.MorseCodePublishing.com/Command

Hiking and Backpacking

For group hiking trips, especially in remote locations, a Baofeng radio is invaluable for keeping team members coordinated over potentially vast distances along the trail. Assign a communications lead to monitor critical frequencies. Group members can provide periodic check-ins regarding their status and location. If anyone strays off course, the radio allows them to be quickly redirected. In case of injuries, it summons help rapidly. The radios also enable coordinating meetups, water/food drops, and pickups. If caught in storms, getting real-time weather alerts from NOAA frequencies can help the group take appropriate precautions (*Best Baofeng Radios*, 2022). In the quiet of the wilderness, the crackle of their Baofeng radios is a comforting reminder of the connection with your group.

Camping

When camping off-the-grid, beyond cell phone service, Baofengs are ideal for keeping fellow campers coordinated, especially after dark when moving around the site. They operate like supercharged walkie-talkies amongst family or friends. Quick communications allow for checking that the perimeter is secured, finding companions who've wandered off, and requesting supplies from others at camp. Monitoring NOAA weather channels provides advance alerts for any severe storms approaching, giving time to batten down gear and take shelter. The radio can summon emergency help quickly should someone face a medical event or threatening wildlife emerge. Or they can be used for just good old entertainment. Tune into local FM stations to enjoy music, news, or radio shows during your outdoor adventures. The versatility and range of a Baofeng radio transforms camping connectivity and safety.

Fishing

On remote fishing trips, a handheld radio provides vital benefits. Frequently scanning weather frequencies keeps the anglers updated on any systems moving in that might cut the excursion short. Voice-activated transmission (VOX) allows quickly radioing others hands-free when reeling in a catch. The emergency tone feature rapidly signals the group if a participant is injured or overboard. And in the worst case of an engine failure at sea, the radio may become a lifeline to call for help when miles from shore and cell coverage. The importance of a Baofeng radio on open water cannot be overstated.

Hunting

For hunters venturing into the wild, a radio is equally crucial. Keeping an ear on NOAA weather channels provides advance notice to exit the field before storms strike. The emergency tone capability rapidly summons aid should a hunter become lost or injured in rough terrain. Voice activation allows quietly transmitting to companions if game is spotted without fumbling for a handset. Some radios even have silent vibration modes to avoid startling wildlife. And if recovering a downed deer or hauling equipment, quick coordination with others via the radio brings essential assistance. For safe and successful hunts, a radio is indispensable gear.

MOTORSPORTS

In the context of motorsports, Baofeng portables genuinely shine. Quick radio communication can save lives at remote race events like desert rallies, or motocross held miles from cellular service. An accident that immobilizes a vehicle can leave racers stranded and vulnerable. A Baofeng allows contacting event coordinators immediately to dispatch medical responders before a situation becomes dire. Or if a race vehicle breaks down out of sight, the radio can guide support crews to the disabled driver. Two-way radios are long-standing saviors in motorsports, and Baofeng models deliver exceptional range and reliability.

In summary, whether deep in the mountains, miles offshore, or racing across the desert, exploring the great outdoors poses risks beyond cell phone service. However, a properly utilized Baofengs provides a flexible emergency lifeline. With their durability, excellent range, and ease of use, these radios empower users with confidence-inspiring communications when adventuring off-grid. So, make a Baofeng mandatory gear for your next foray beyond the pavement. Their importance is immeasurable if an emergency arises.

USING YOUR RADIO FOR EVENT COORDINATION

For events large and small, Baofeng radios provide an indispensable tool for communication and coordination between staff spread over a wide area. Let's explore typical use cases where these versatile radios shine for events.

Neighborhood Watch Program

In a quiet neighborhood where the most exciting event is the weekly book club meeting, the arrival of a Baofeng radio sparked a small revolution. When a spate of burglaries shook the peace of this tranquil community, the residents decided to take action. They formed a neighborhood watch program, and their Baofeng radios became their most trusted tool.

The radios facilitated real-time communication among residents, from coordinating patrols to reporting suspicious activities. The local frequency became a hub of community coordination, a testament to the spirit of camaraderie and vigilance. The burglaries dropped, and the peace was restored. The residents continued to use their Baofeng radios, not just for the neighborhood watch but for coordinating community events, sharing local news, and sometimes, just for a friendly chat.

Local Ham Radio Club

In a bustling city, a group of radio enthusiasts found solace in the airwaves. They formed a local ham radio club, and their Baofeng radios were their tickets to this exclusive fellowship. The club members used their radios to learn from each other, share their experiences, and revel in the joy of radio communication.

They organized 'fox hunts,' in which a transmitter is hidden, and members use their Baofeng radios to locate it. They also conduct workshops to help new members get their licenses and master their radios. They even set up a station to communicate with astronauts aboard the International Space Station! For this radio club, their Baofeng radios are more than just devices; they are keys to a world of endless learning and exploration.

Local Community Events

For small local events like school fundraisers, fun runs, and street fairs, handhelds allow rapid communication between event organizers and volunteers. Channels can be pre-programmed for key roles —ticket sales, parking coordination, setup/teardown crews, etc. Requests can be made over the radios if extra staffing is needed in the ticket booth or parking. For widely dispersed events, radio contact keeps everything running smoothly.

Festivals and Concerts

At large music festivals spanning extensive grounds, reliable communication is vital yet challenging. Cell networks quickly become overloaded with thousands of attendees, so radio is essential. Baofengs allow staff to remain connected across the site. If the water station needs supplies, the medical tent requires an ambulance, or gates require more security—urgent assistance can be summoned quickly over the radio. Channels help sort important communications, like coordinating artists and VIPs discreetly via dedicated security channels.

Marathons and Races

Marathons and races pose similar challenges, with thousands of participants spread over a lengthy course. Spotters and medical personnel along the route can radio in any developing situation. Race officials communicate progress, lap counts, and pack positions. Baofengs prove perfect for the mobile nature of races, keeping coordinators connected across long distances. Reliable communication is crucial if any emergencies arise.

Rural Events

Cell coverage is often nonexistent in remote rural locations. However, Baofengs provide clear communication regardless of terrain or cell congestion. Radio is the only way staff can efficiently coordinate and respond to issues for outdoor events like cattle shows, air shows, or motocross rallies in the sparse country. Emergency services also utilize the radio frequencies to provide urgent medical care when required.

Corporate Events

Seamless coordination ensures perfect execution at highly choreographed corporate events and conferences with key executives in attendance. Baofengs enable staff and technical directors to flawlessly sync speeches, videos, and lighting cues. Security personnel can also discreetly communicate threats if issues arise. These robust radios operate reliably even in large buildings and exhibition halls.

Transportation Support

Radios are indispensable at transportation hubs like bus depots and airports. Staff can use them to redirect passengers, dispatch buses efficiently during peak loads, report mechanical problems, and request maintenance crews rapidly. Radio contact between drivers and dispatchers keeps operations moving smoothly for widely dispersed parking lots and transit centers. Reliable communication is essential for airports and transit systems.

EMERGENCY/CONTINGENCY PLANNING

While festivals and events are joyous, emergencies can arise from weather, accidents, or other unforeseen issues. Event planners prepare contingency protocols using radio to coordinate urgent responses. Channels are pre-allocated for emergency services, medical teams, security, and evacuation decision-makers. Radios allow swift deployment of contingency plans even if normal communications channels become overloaded or fail.

In summary, smooth coordination underlies every successful event, whether an intimate gathering or a massive concert. Baofeng radios enable robust and reliable contact between planners and staff, even in remote locales or congested environments. For rapid response to changing needs, contingencies, and safety, professional two-way radios prove indispensable tools behind the scenes of events worldwide.

Off-Grid and Survival Situations

Baofeng radios truly shine during crises when standard infrastructure fails. Being able to communicate off-grid can save lives. Let's explore some scenarios where dependable handheld radios prove invaluable for resilience.

Wilderness Expeditions & Survival

Envision a scenario where you're hiking, injured, and without a cell signal. In such a moment, a Baofeng radio enables you to reach Search and Rescue teams or connect with your group for guidance. For outdoor adventurers, a reliable handheld radio like the Baofeng is an essential safety tool, providing communication where cell networks don't.

While satellite phones are helpful in remote areas, affordable, portable radios like Baofengs are vital for team communication and emergency coordination. They offer advanced communication capabilities, ensuring you're prepared for any situation. Regular training with your Baofeng makes you a valuable asset to your community in emergencies, turning you from a mere user into a crucial part of a larger support network.

Severe Weather Events

When violent storms, flooding, or fires disrupt everyday communications, amateur radio fills the void. Hams band together to relay status reports and call for assistance. During Hurricane Katrina and many other disasters, ham radio was the only consistently functioning communication system. Baofengs and other handhelds are perfect for rapidly deploying ad-hoc networks. Your skills could save lives by summoning rescues for stranded victims.

HOMESTEADING AND OFF-GRID LIVING

For modern homesteaders living sustainably off-grid, especially in remote regions, Baofeng radios have become a crucial tool, offering a vital link to surrounding communities. These radios enable the sharing of news and organizing assistance. Still, they are also pivotal in monitoring threats and facilitating bartering, providing a reliable form of communication that is indispensable in areas where landline phones are vulnerable to weather disruptions.

Baofeng handhelds and base stations offer an affordable, hardened redundancy that off-grid groups like preppers widely recognize as essential gear for self-reliance. When traditional communication networks are down in emergencies, Baofeng radios serve as a critical lifeline, allowing homesteaders to receive updates on natural disasters, seek help, and coordinate community responses. They also facilitate regular check-ins with neighbors, enhancing local networking and community building, essential for remote living.

One of these radios' most crucial uses is monitoring weather and environmental conditions. Access to real-time weather updates and alerts is vital for planning and preparedness, especially in areas prone to severe weather conditions. For those involved in agriculture, Baofeng radios enable efficient coordination across large regions, which is essential for coordinating harvests and managing livestock.

Additionally, these radios play a significant role in security and surveillance, providing a means for homesteaders to communicate instantly in case of suspicious activities or threats. They are also used as a tool for educational purposes, facilitating the exchange of information and skills crucial for sustainable living.

The technical advantages of Baofeng radios are particularly suited to the needs of homesteaders. Known for their durability and reliability, they are well-equipped to handle the demanding conditions of off-grid living. With extended battery options, these radios ensure communication lines stay open, even during power outages. Users can also program specific frequencies for private group communication, enhancing privacy and reducing interference.

Furthermore, the range of these radios can be extended with external antennas and repeaters, a vital feature for use in sprawling rural areas. Their affordability makes them a cost-effective communication solution, lowering the entry barrier for homesteaders and preppers. Additionally, Baofeng radios are user-friendly and require minimal technical knowledge for basic operations, ensuring they are accessible to many users.

Baofeng radios are more than just communication tools for the modern homesteader and prepper; they are a gateway to a self-reliant, resilient lifestyle. Offering versatility, reliability, and accessibility, they are a wise investment for anyone seeking to establish a sustainable, connected homestead. Their practical applications in emergency, agricultural, and community settings and their technical benefits make them indispensable to off-grid living.

SOCIETAL COLLAPSE

In a societal collapse scenario, Baofeng radios emerge as invaluable tools for survival and rebuilding communities. Traditional communication infrastructures will likely be disrupted or utterly nonfunctional during such tumultuous times. In this context, Baofeng radios offer a reliable means for maintaining communication, which is crucial for safety, coordination, and community resilience.

Baofeng handhelds and base stations have become pivotal in establishing emergency communication networks. They enable individuals and groups to share critical information, such as safety alerts and resource locations, and coordinate group efforts in response to ongoing challenges. When timely information can mean the difference between safety and danger, these radios provide a lifeline.

The radios' versatility in accessing a wide range of frequencies is particularly beneficial. Users can tap into various channels to gather news, connect with others, or set up localized networks for community defense and mutual aid. Their ability to operate on VHF and UHF frequencies allows for local and extended-range communications, which is essential in a fragmented society.

Regarding logistics and resource management, Baofeng radios facilitate organizing essential services such as food distribution, medical aid, and shelter management. They enable groups to coordinate scavenging missions, communicate about safe and dangerous areas, and organize trading or bartering systems necessary for survival.

From a security standpoint, these radios are crucial for setting up surveillance and alert systems. In a societal collapse, safeguarding communities against various threats is paramount. Baofeng radios provide a means for constant vigilance and quick response, enhancing collective security.

The rugged and durable nature of Baofeng radios ensures their reliability in harsh conditions. Coupled with their long battery life and the option for rechargeable batteries or alternative charging methods (like solar chargers), they are well-suited for prolonged use in scenarios where conventional power sources are unavailable.

Moreover, their affordability and ease of use make them accessible to a broader population. This accessibility is crucial in a societal collapse, where complex or expensive communication solutions would be impractical or unattainable for most people.

In the face of societal collapse, Baofeng radios are essential for survival, organization, and community resilience. Their ability to facilitate reliable communication, coordination, and security in a world where traditional infrastructures have failed makes them indispensable in navigating and adapting to the new challenges of a drastically changed world. Their technical robustness, user-friendliness, and affordability ensure they can be widely adopted to rebuild and reconnect fragmented communities.

Emergency Radio Groups

Should catastrophes ever disrupt normal society, resilient communication will prove critical. Amateur radio groups like ARES, RACES, and CERT drill extensively for these emergency scenarios.

- **ARES (Amateur Radio Emergency Service)**: ARES is a program run by the American Radio Relay League (ARRL). It consists of licensed amateur radio operators who volunteer to support emergency communications during disasters and public service events. ARES members train to offer their skills in radio communications to assist local authorities and emergency response agencies when regular communication systems fail or are overloaded.
- **RACES (Radio Amateur Civil Emergency Service)**: RACES is a public service provided by amateur radio operators, established by the Federal Emergency Management Agency (FEMA) and the Federal Communications Commission (FCC). In a national emergency, RACES volunteers provide essential communications and are activated by local, county, or state emergency management agencies. Unlike ARES, which can be activated for a broader range of public service

events, RACES is strictly limited to emergency or disaster
situations where everyday communications are unavailable.

Rapidly deploying radio nets using optimized equipment like Baofeng portables enables community coordination for safety, news, and resources. Hams become the communication backbone, recovering order from chaos. Should the unthinkable occur, dedicated preppers equip themselves with radios, knowledge, and training to assist others.

Your Baofeng radio is more than just a radio; it's a versatile tool, ready to assist you in various aspects of your daily life. So, whether you're chatting with your neighbor, tracking a storm, exploring the great outdoors, or volunteering at a local event, your Baofeng radio will be right there with you, enhancing your experiences and ensuring you're never truly alone.

Now that you've mastered operation, features, and real-world applications for your Baofeng radio, let's focus on keeping it running smoothly long-term. The key to reliability is proper maintenance and knowing how to troubleshoot issues when they arise.

In our final chapter, we'll cover essential topics like care, cleaning, accessories, and troubleshooting common problems with your radio. Proper upkeep will ensure your Baofeng provides years of flawless performance when needed.

10

FIELD GUIDE - MAINTENANCE, CARE, AND PROBLEM SOLVING

I n the world of prepping and preparedness, your Baofeng radio is not just a tool but a critical asset for communication in times of crisis. Understanding proper maintenance and troubleshooting is essential to ensure its reliability when you need it most.

This chapter is dedicated to empowering you with the knowledge to keep your Baofeng radio in top working condition. You'll learn the ins and outs of regular upkeep so your radio remains operational in various conditions. We'll also delve into common issues and their solutions, equipping you with the skills to diagnose and fix problems quickly.

MAINTAINING YOUR BAOFENG RADIO

Proper maintenance and care ensure your Baofeng radio delivers years of flawless performance. With the appropriate upkeep, these radios can serve reliably even in harsh conditions. We will cover essential maintenance best practices, including cleaning, storage, accessories care, and periodic refresh procedures.

Exterior Cleaning

Maintaining the exterior of your Baofeng radio is crucial for preserving its appearance and functionality. Over time, exposure to dirt, dust, and grime can lead to various issues, including:

- Blocked speaker holes, which can diminish audio quality.
- Buttons becoming stiff or sticky due to grime accumulation.
- Potential abrasion damage to the display if dust is left uncleaned.

To ensure your radio remains in top condition, follow these cleaning guidelines:

- Always turn off the radio and remove the battery before cleaning.
- Use a lightly dampened microfiber cloth with a bit of mild detergent. Be sure to wring out any excess water from the fabric.
- Gently clean all external surfaces, focusing on crevices, buttons, and the display area.
- Do not introduce moisture into sensitive areas such as the battery compartment, speaker holes, and microphone holes.
- After cleaning, use a dry cloth to remove moisture and let the battery air dry completely before reassembling.
- For tougher dirt, a small amount of isopropyl alcohol can be effective, but ensure it doesn't seep into the radio's internal components.
- Do not use harsh chemicals, solvents, bleach, or abrasive cleaners, as these can damage the radio.

Regular external cleaning will keep your radio looking new and prevent performance issues related to dirt and dust accumulation.

Antenna Care

Proper care of your radio's antenna is essential, as it is crucial in signal transmission and reception. Regular antenna maintenance ensures it functions at its best, providing optimal range and clear communication.

When inspecting your antenna, particularly for Baofeng models like the UV-5R, pay attention to the following:

- Check the SMA male connector for any signs of oxidation or carbon buildup, which can occur due to exposure or aging. If present, gently clean the connector using isopropyl alcohol.
- Examine the antenna's rubber gasket. If it's cracked or deformed, replace it to maintain the antenna's water resistance.
- Check the metal antenna shaft for any physical damage, such as cuts, nicks, or dents. Damage like this can degrade the antenna's performance, so it's best to replace it in these cases.
- Inspect the antenna tip for carbon deposits, which might result from arcing. Carefully remove these using fine sandpaper.

Battery Maintenance

Proper battery care practices can significantly extend the battery's runtime and overall lifecycle. Here's how to get the most out of your Baofeng radio's battery:

- Always use manufacturer-approved Li-ion battery packs. Refrain from using cheaper, non-approved clones to ensure safety and efficiency.
- Regularly inspect the battery for any signs of damage, such as swelling, and safely dispose of any compromised batteries.

- Keep the contact terminals clean to prevent corrosion. Gently wipe them with alcohol periodically.
- To prolong battery life, avoid completely draining the battery before recharging.
- When not using the battery, remove it from the radio and store it separately.
- Charge the battery in moderate temperatures. Avoid charging in extremely hot or cold environments like an uninsulated garage or shed.
- Occasionally, perform a complete discharge and recharge cycle. This helps to calibrate the battery and can maintain its capacity.

Storage Best Practices

Storing your Baofeng radio correctly when it's not in use is crucial for preserving its condition and extending its service life. Proper storage methods can prevent damage and maintain the radio and its batteries in optimal condition (*Baofeng Cheat Sheet,* n.d.). Here are essential practices to follow:

- Store the radio in a cool, dry place, away from extreme temperatures. High-heat areas, like vehicles, should be avoided.
- Prefer indoor storage to protect the radio from dust and moisture, which can be harmful if exposed to temperature or moisture changes, for example, in a shed or non-temperature control garage.
- Avoid prolonged exposure to direct sunlight, as it can damage the radio's surfaces and compromise its seals.
- Store the radio upright, in its intended orientation, rather than laying it flat. This helps prevent potential damage to its buttons and other components.

- For long-term storage, remove the batteries from the radio. This practice helps maintain the battery's health and prevents potential leakage or damage.

Waterproofing Considerations

Ensuring your Baofeng radio's protection against water is essential, especially since most models are not inherently designed to be waterproof. Even though some high-end models may offer water resistance, taking additional measures to safeguard your device in wet conditions is crucial. Here are some effective waterproofing strategies:

- Utilize a waterproof carry case or bag when operating in moist environments. Many cases are designed to allow access to the radio's buttons and display while providing protection.
- Consider applying waterproofing sprays to the radio's exterior for temporary defense against light rain or splashes. Remember to reapply the spray following any exposure to water.
- After using the radio near or in saltwater, such as during boating or coastal activities, rinse it with fresh water to prevent corrosion damage.
- Avoid submerging the radio in water unless enclosed in a case specifically designed for immersion.

By taking these precautions, you can significantly extend the life and maintain the functionality of your Baofeng radio, even in challenging environmental conditions.

PROGRAMMING BACKUPS

Maintaining the safety and integrity of your radio's programmed frequencies and settings is crucial, as they are stored in internal memory that could be susceptible to corruption or loss. Regular backups are an intelligent way to safeguard this vital data. Here's how you can efficiently back up and maintain the settings of your radio:

Backup Process:

- Connect your radio to a computer using the appropriate cable.
- Utilize software like CHIRP to read and save the current configuration of memory channels and settings.
- Consider storing backups in multiple locations, such as cloud storage or an external hard drive, for extra security.

Verifying Backup Data:

- After backing up the data, carefully review the saved channel information to ensure no errors or missing details.
- Regular checks are essential to confirm the integrity and completeness of your backup.

Maintaining Backup Data:

- It's advisable to refresh your backup data annually or whenever you experience issues like dropped channels or other anomalies.
- Keeping your backup current ensures you have saved the latest settings and configurations.

Restoration:

- In the event of data corruption or loss, having a recent backup allows you to swiftly restore your radio to its previous, optimal configuration.
- This quick restoration is precious in field operations where reliability is paramount.

Adhering to these best practices for backup and maintenance will ensure that your radio's settings and frequencies are well-protected and easily recoverable. This proactive approach will save time in troubleshooting and ensure your radio is always ready for use with your preferred configurations.

ACCESSORY MAINTENANCE AND CARE

Maintaining and caring for your chargers ensures ongoing performance and reliability. Here's how you can effectively maintain these critical components:

Charger Maintenance:

- Regularly inspect the charger contacts for any build-up of grime or oxidation. If necessary, clean them using alcohol to ensure efficient charging.
- Avoid leaving the radio on the charger continuously after fully charging. Overcharging can significantly reduce the battery's lifespan.
- If the charger is damaged or its efficiency is reduced, replace it with an original equipment manufacturer (OEM) product to ensure compatibility and safety.

Drop-in Desktop Charging Station: A drop-in desktop multi-bay charging station can be incredibly convenient for those who frequently use their radios. Here are its key benefits:

- Allows for simultaneous charging of multiple radios and spare batteries, enhancing efficiency.
- Eliminates the need for constantly connecting and disconnecting individual chargers.
- The cradles in the station ensure secure docking and maintain proper contact for consistent charging.

- Indicator lights provide a quick and easy visual of the charging status of each bay. Such a charging station is a time-saver for heavy users, keeping batteries fully charged and ready for use. Common models include 6 or 12 bays catering to different usage needs.

ADVANCED TROUBLESHOOTING

Software Diagnosis

Advanced troubleshooting of your Baofeng radio can be conducted with programming software like CHIRP, especially for users comfortable with more technical aspects. This software offers a range of diagnostic tools and capabilities that can aid in identifying and resolving complex issues:

- Reading Radio Data: By accessing the radio's internal data, CHIRP can help identify issues like memory corruption or incorrectly stored frequencies, which can be crucial in pinpointing problems that aren't immediately obvious.
- Error Logs and Debug Modes: CHIRP often provides access to error logs and debug modes, offering more profound insight into any faults or anomalies your radio might be experiencing.
- Monitoring Signals: The software can monitor transmitting and receiving signals in real-time, which is invaluable for on-the-spot debugging and ensuring optimal radio performance.
- Factory Reset: In cases where issues arise from incorrect custom programming, CHIRP can facilitate resetting the radio to its factory settings, often resolving such glitches.

For most users, basic troubleshooting methods are sufficient. However, programming software unlocks a higher diagnostic and analysis potential, allowing advanced users to delve deeper into their radio's functionality and address more complex issues. This advanced approach can be incredibly beneficial for maintaining the reliability and performance of your Baofeng.

Full Factory Reset

A full factory reset on your Baofeng radio can be necessary if you encounter persistent glitches or anomalies that standard troubleshooting doesn't fix. Here's a step-by-step guide to ensure a successful reset:

1. Backup Radio Memory:

- If you want to retain your channel programming, back it up using computer software. This step ensures that your channel information is preserved for later restoration.

2. Initiating Full Factory Reset:

- Press the MENU button on your Baofeng radio.
- Use the arrow keys to navigate to the 'RESET' menu option. On most versions, it will be the very last Menu option.
- Select the option for a full reset and confirm the action when prompted.

3. Reprogramming the Radio:

- After the reset, reprogram your radio from scratch. Re-enter any saved channels and settings if necessary.

4. Testing Post-Reset:

- Thoroughly test the radio to ensure that all issues have been resolved. Monitor its performance across all features before putting it back into regular use.

A full factory reset is often a last resort but can effectively resolve severe misconfigurations or programming bugs. This careful approach will help restore your radio to its optimal functioning state.

KEY PREVENTATIVE MAINTENANCE SCHEDULE

Maintaining your radio in top condition requires a proactive approach. Here's an easy-to-follow preventative maintenance schedule to help you stay ahead of potential issues:

Daily Checks Before Use:

- Conduct a visual inspection of your radio for any signs of damage.
- Ensure the battery is adequately charged and the charger is ready.

Monthly Care After Heavy Field Use:

- Give the radio a thorough external clean, focusing on the body, buttons, display, and connectors.
- Inspect the antenna connection and its gasket for any wear or damage.
- Clean the battery terminals if you notice any buildup.

Annual Maintenance:

- Perform a complete functionality check and a range test to verify the radio's performance.
- Back up and refresh your memory channel data to safeguard your settings.
- Go through a complete battery discharge and recharge cycle to optimize battery health.

As Required:

- Apply waterproofing treatments if you frequently use the radio in wet conditions.
- Update your programming software to the latest version for enhanced functionality.

By dedicating time to these primary care routines, you can identify and address minor issues before they escalate. Developing expertise in these maintenance procedures is a surefire way to prevent more significant, more complex problems down the line. Remember, even the most reliable equipment requires consistent care—don't overlook the importance of routine maintenance for seamless operation.

UNDERSTANDING & TROUBLESHOOTING COMMON ISSUES

Understanding and troubleshooting common issues with your Baofeng can enhance your user experience and ensure longevity and preparedness. While technical glitches can be frustrating, most problems have straightforward solutions. This section will guide you through the most common issues users face with their radios and provide practical fixes.

Radio Not Powering On The inability to power on can be highly disruptive, but common causes include:

- Dead Battery: Verify the battery's charge with a spare pack. New batteries may need charging.
- Faulty Power Button: Inspect for dirty or damaged contacts if the battery is functional.
- Blown Fuse: Use a multimeter to check for continuity.
- Charging Circuit Issues: Faulty components can inhibit charging and power from the battery.

- Mainboard Failure: Internal regulator or power control issues may require professional repair.

Poor or No Reception: Reception problems often stem from antenna issues:

- Loose or Detached Antenna: Ensure it's fully screwed on for proper contact.
- Damaged Antenna: Inspect for physical damage and replace if necessary.
- Damaged Center Pin: Check for damage or corrosion in the antenna connector.
- Defective Receiver: The receiver module may need repair if the antenna is OK.

Battery Issues: Common battery-related problems include:

- Rapid Drainage: Regularly discharge and recharge to maintain battery health.
- Dirty or Faulty Contacts: Clean contacts to prevent power issues.
- Physical Damage: Check for swelling or leaks and replace damaged batteries.
- Counterfeit Batteries: Use only branded batteries for reliability.

Keypad and Button Issues: Wear and tear can lead to unresponsive buttons:

- Contamination: Clean around buttons to remove dirt.
- Stuck Contacts: Inspect and gently adjust the interior rubber membrane.
- Physical Damage: Repair or replace damaged components.

Sudden Loss of Transmit Ability: If transmit capability is suddenly lost, consider the following:

- Antenna Mismatch: Ensure antenna impedance compatibility.
- Faulty Power Amplifier: Test and replace if necessary.
- Internal Wiring Issues: Inspect coaxial cables for damage.
- Mainboard Problems: Check transmit-related components for faults.

Distorted or Reduced Volume: Audio issues can be caused by:

- Obstructed Speaker: Clean any blockages carefully.
- Low Battery Voltage: Replace with a fully charged battery.
- Damaged Speaker: Replace speakers with physical damage.
- Bad Solder Joints: Professional repair may be required for board issues.

Programming Difficulties: Challenges in programming can arise from:

- Outdated Software: Use the latest version of programming software.
- Cable Issues: Check for faulty connections and use quality cables.
- User Error: Review programming steps and data.
- Memory Corruption: Reset the radio to factory settings if necessary.

Repair Options: Consider various repair options before replacing your radio:

- Manufacturer Warranty: Utilize the warranty for eligible repairs.
- Local Repair Shops: Obtain an estimate for economical fixes.

- DIY Repairs: Simple replacements like antennas or batteries can be user-performed.
- Mail-In Service: For complex issues, consider specialized repair services.

Avoiding Future Problems: Preventive measures can help avoid common issues:

- Careful Handling: Use cases and straps to prevent physical damage.
- Moisture Prevention: Keep the radio dry, especially if it's not water-resistant.
- Proper Storage: Store in protective cases away from heavy objects.
- Regular Inspection: Check for wear and tear and replace parts proactively.
- Accessory Maintenance: Use only compatible, branded accessories.

By understanding these common issues and their solutions, you can swiftly address problems and ensure your Baofeng radio remains a reliable communication tool. Regular care and preventive measures will also contribute to its durability and functionality.

SEEKING TROUBLESHOOTING HELP

When facing complex technical issues with your Baofeng that surpasses your expertise or when special diagnostic tools are required, seeking external help can be invaluable. Here's how you can find assistance to address your radio problems effectively:

- **User Forums:** Online communities like the Baofeng subreddit or RadioReference forums are great resources. Here, fellow radio enthusiasts share troubleshooting tips and solutions. You can search for existing threads addressing similar issues or post your queries for personalized advice.

- **Local Repair Shops:** For intricate component-level problems, local electronics repair shops can be a lifesaver. They have specialized tools, like oscilloscopes, to diagnose issues precisely. Often, the cost of repairs is more economical than replacing your radio.
- **Technical Manuals:** Detailed service manuals designed for technicians offer comprehensive information on your radio. These manuals include schematics, standard measurement values, and guidance on disassembling your device for advanced troubleshooting.
- **Manufacturer Support:** If your radio is under warranty, contacting Baofeng's support can be beneficial. Be prepared to provide details like the model, a summary of the issue, and any troubleshooting steps you've already attempted.
- **YouTube Tutorials:** Numerous video tutorials cover common repairs and troubleshooting for specific Baofeng models. These visual guides can walk you through the process step-by-step, making complex fixes more approachable.

If you struggle with persistent radio issues, remember that a wealth of knowledge and assistance is available. From the collective wisdom of online forums and instructional videos to the expertise of professional repair services, these resources can help you effectively resolve even the most challenging problems. Don't hesitate to tap into these options to get your Baofeng radio back in optimal working condition.

CASE STUDIES: TROUBLESHOOTING IN ACTION

Overcoming Reception Issues in a Dense Forest

Consider a group of outdoor enthusiasts exploring the dense, verdant woods of the Pacific Northwest. Their Baofengs, strapped to their backpacks, are their lifelines, keeping them connected as they wander through the wilderness. But as they delve deeper into the forest, they notice a problem. The lush canopy overhead, though beautifully serene, is causing reception issues on their radios.

Instead of succumbing to frustration, they take a pragmatic approach. They recall the troubleshooting tips they've learned and put them into action. They check their antennas, ensuring they are properly connected and not damaged. They tweak the squelch settings on their radios, filtering out background noise and improving signal clarity. Yet, the dense foliage remains a formidable adversary.

One of the group members, an experienced radio operator, suggests a simple yet effective solution - elevation. They find a clearing with a tall tree. One of them, an adept climber, volunteers to climb up, radio in hand. As she ascends, the reception starts to clear. The higher she climbs, the clearer the signal becomes. From her elevated perch, she is able to relay information to the rest of the group and successfully guide them through their woodland adventure.

Alternatively, this group could have packed a portable antenna system. This could be a lightweight, portable antenna specifically designed for field operations. An example would be a collapsible or roll-up J-pole antenna made from lightweight materials. This type of antenna can be easily deployed, and offers improved reception over the standard rubber duck antennas typically found on handheld radios.

In addition, knowing if there are repeaters in the area is a viable option that should be tried.

Solving Battery Drain During a Power Outage

In a small town struck by a severe storm, the power grid has failed, plunging the townsfolk into darkness. Their Baofeng radios are their only means of connecting with the outside world. But as the hours turn into days, they notice their radio batteries draining rapidly.

One resident, a seasoned radio operator, steps up to tackle the issue. He advises the townsfolk to turn off unnecessary features on their radios, such as the flashlight and the backlight on the display. He then teaches them how to set their radios to Battery Save Mode, which conserves power when the radio isn't actively receiving a signal.

Despite these measures, the battery drain continues. That's when the radio operator pulls out his secret weapon - a car charger adapter for Baofeng radios. Using the power from a car battery, he is able to recharge the radio batteries, ensuring they stay operational until power is restored. His quick thinking and practical knowledge save the day, keeping the lines of communication open during this crucial time.

This is another perfect example of where your battery bank will keep you safe!

Fixing Programming Errors Before a Community Event

In a vibrant community preparing for an annual fair, Baofeng radios are buzzing with activity. The event organizers use the radios to coordinate preparations, assign tasks, and keep everyone updated. But as the event day approaches, they encounter a problem. Their attempts to program new channels into their radios for different event teams are met with errors and frustrations.

Noticing their struggle, a member of the community steps forward. She's a radio enthusiast with a knack for troubleshooting. She looks at the radios and identifies the problem - a simple programming error. The organizers were trying to save the new channels in already occupied slots.

She guides them through deleting the existing channels before programming the new ones. She also shows them how to name the new channels according to their respective teams, making it easier for everyone. Thanks to her, the programming errors are quickly resolved, and the community event is a resounding success.

Addressing Audio Problems in a Search and Rescue Operation

A search and rescue team is on a mission high in the mountains. Their Baofeng radios are their lifelines, connecting them and their base camp. But as they ascend higher, they start experiencing audio problems on their radios. The volume is too low, and the sound is distorted, making communication difficult.

One of the team members, an experienced radio operator, decides to troubleshoot the issue. He checks the volume settings on their radios, ensuring they are turned up. He inspects the speakers for any blockage or damage but finds none.

Then, he remembers a trick he learned. He suggests using an external speaker/mic. The team has a few in their gear meant to facilitate hands-free communication. They connect the speaker/mics to their radios, and instantly, the audio is louder and clearer.

With the audio problems resolved, the team can communicate effectively, coordinate their efforts, and complete their rescue mission.

Navigating the airwaves with your Baofeng is like embarking on an exciting adventure. Each transmission is a step forward, each frequency a new path to explore. And like any adventure, there might be challenges along the way. But with every challenge comes an opportunity to learn, grow, and become a better radio operator. So, whether you're scaling the heights of a mountain, navigating the depths of a forest, powering through a blackout, or bringing a community together, your Baofeng radio will be right there with you, ready to conquer any challenge that comes your way.

Your Role in the Radio Community

Radio is only as good as the community behind it—so let's build that community together.

Simply by sharing your honest opinion of this book and a little about what you found here, you'll help new readers uncover this vital survival information.

RATE US!

Thank you so much for your support. May you always be well prepared!

CONCLUSION

WRAPPING UP: THE JOURNEY SO FAR

As our exploration of the Baofeng radio universe draws to a close, it's time to reflect on the incredible journey we've embarked upon together. Our voyage has been enriching and enlightening, from the fundamental understanding of radio anatomy to mastering programming techniques and recognizing these devices' indispensable role in everyday and emergency scenarios.

We began this adventure with a simple goal: to break through the complex world of Baofeng radios. What we uncovered was far more profound—a realization of these radios' immense impact on our lives and safety. Beyond their technical capabilities, Baofeng radios have emerged as powerful instruments of connection and change, fostering community ties and serving as critical lifelines in times of need and disaster.

These devices remind us that despite the many barriers in our world, the airwaves offer a universal platform for connection and understanding. They empower us as radio operators and active participants in a global dialogue, even when our phones fail us.

KEEP LEARNING: JOIN THE BAOFENG COMMUNITY

Our adventure doesn't conclude here. The Baofeng radio world is vast and continuously evolving. I encourage you to delve deeper, seek new knowledge, and continually expand your understanding. A wealth of resources is available—online communities, user groups, and advanced literature—that can further refine your proficiency and enjoyment of Baofeng radios.

I warmly invite you to become an active member of the vibrant Baofeng community. Share your experiences, learn from fellow enthusiasts, and contribute to a global network of radio aficionados. Every transmission sent and received is a step towards thoroughly preparing your communications.

SIGNING OFF

In these pages, we aimed to transform beginners into informed, skilled guerilla Baofeng users. From selecting the right model to mastering initial setups and advanced operations, our goal was to simplify and clarify, turning complexity into confidence.

We've tackled key aspects like effective programming, achieving clear communication, navigating legalities, and delving into advanced features, equipping you for a fulfilling long-term engagement with your radio and the preparedness to survive.

Remember, this book is but an introduction to a lifelong journey. Continue to engage with the community, participate in forums, drill, attend events, and train with your radio regularly. Challenges are growth opportunities, and proficiency comes with practice.

It has been an honor and a pleasure to introduce you to the fascinating world of Baofeng radios. I hope you feel empowered and curious enough to embark on or continue your journey. The world of two-way radio communication is vast and waiting for you.

Your real adventure begins now as a guerilla Baofeng user who has broken through the complexity secured their communications, and become prepared for disaster utilizing prepper tactics!

- QRV?

BONUS SECTION: A BEGINNER'S ESSENTIAL GUIDE – YOUR NEXT STEPS

Congratulations! You made it through the book, but I'm guessing you still have questions. You still aren't 100% sure what to do next. That's exactly how I felt at this same stage. I wanted just a little more direction. So here you go! This cheat sheet provides a simplified roadmap to get you started on the right foot. Follow these steps to start moving from the beginning guerilla to Baofeng master.

STEP 1: PURCHASE THE BAOFENG UV-5R RADIO

The UV-5R is the classic starter model, renowned for its affordability, versatility, and tremendous user community support. I recommend buying it from an authorized dealer like BaoFengTech.com or Amazon to guarantee authenticity.

Some key benefits of the UV-5R for beginners (*Best Baofeng Radios*, 2022; Tobias, 2015):

- Dual-band support for VHF and UHF frequencies right out of the box. This allows you to access a wide range of amateur radio bands as you get started.

- Highly customizable via easy PC programming with CHIRP software. The UV-5R was one of the first accessible handhelds that opened programming for beginners through a computer.
- There is a vast online knowledge base and active user forums. As the most popular starter radio, every aspect of the UV-5R has been discussed and documented by the global community. This community provides troubleshooting support and hacks/mods to customize the radio as your skills progress.
- Key Benefit: Affordably priced between $25–50 depending on accessories bundled. This low cost allows anyone to get started without a significant investment. Which is a very important and often overlooked aspect for a beginner.

While imperfect, the UV-5R is built to last and, with proper care, will serve you well for years. Overall, it's the ideal starter radio.

STEP 2: OBTAIN YOUR GMRS LICENSE

While the UV-5R can operate license-free on FRS frequencies, I highly recommend getting the GMRS license. This unlocks higher transmit power on GMRS frequencies for enhanced range.

Key features of this license:

- No test needed to obtain.
- Minimal fee of $35, which is valid for ten years.
- You will receive a GMRS call sign to use in your communications.

The minimal investment is well worth it. Your UV-5R can transmit up to 5 watts on GMRS repeaters and frequencies for better reach. This opens up your communication versatility—you can still use FRS for short-range neighborhood watch coordination, but switch to GMRS when you need more distance, like on hikes or remote events.

Follow this Process to Obtain Your GMRS License

- Obtain your 10-digit FCC Registration Number (FRN) by clicking on 'REGISTER' here: https://apps.fcc.gov/cores/userLogin.do
- Once you have your FRN, login to the FCC License Manager website: https://wireless2.fcc.gov/UlsEntry/licManager/login.jsp
- Click on the 'Apply for a New License' button.
- Then, select 'ZA – General Mobile Radio Service (GMRS).' It begins with a 'z,' so it will be near the bottom of the dropdown list.
- Complete the application and then submit your payment.
- Your new call sign will be assigned to you.

STEP 3: FIND FREQUENCIES RELEVANT TO YOU

Start with the basic FRS/GMRS national calling channels for interoperability with other users - program in all the 22 designated frequencies in the UHF band.

With your GMRS Callsign, register with *myGMRS.com* and find local repeaters in your area: https://mygmrs.com/map

Finally, if you haven't done it already, find and program your local NOAA weather radio (refer to Chapter 4). This frequency provides weather and emergency alerts for your local area.

This gives you a basic group of frequencies, from which you will expand over time.

STEP 4: DOWNLOAD CHIRP

Download It Here: https://chirp.danplanet.com/projects/chirp/wiki/Home

Use CHIRP to program the frequencies from step #3. Refer to Chapter 4 for step-by-step instructions.

STEP 5: THE ESSENTIAL ACCESSORIES

Start with an extra battery for all-day operation, an aftermarket antenna upgrade for increased range, and a lapel microphone for convenient hands-free use. Quality accessories dramatically improve your experience (*Best Baofeng Radios*, 2022).

Specific recommendations

- Programming Cable: If your radio bundle didn't include a programming cable, I recommend buying one. Ensure it works with your model type.
- Extra battery: An 1800+ mAh high-capacity Li-ion replacement battery provides longer runtime. Avoid cheap clones. With batteries, always remember '2-is-1 and 1-is-none'. If you lose your battery, you also lose your radio. That's why I always say '3-is-for-me.'
- Aftermarket antenna: A compact Nagoya NA-771 15.6-inch upgraded antenna gives 2–3x the range over the stock antenna.
- Lapel microphone: Clip the mic near your mouth for clear voice pickup and hands-free operation. Look for models with an earphone.
- Carry case: A sturdy belt clip case protects your radio and keeps it handy while out and about.

These affordable add-ons truly elevate the usefulness of your UV-5R for outdoor activities and emergencies.

STEP 6: JOIN A BAOFENG AND HAM GROUP

Connect with the passionate Baofeng user community. Stick with the basics. Find one local in-person group and then join a few Facebook groups. They provide troubleshooting support, inspiration for expanded learning, accessory recommendations, and a group of like-minded individuals.

Suggestions

- In-Person Local Clubs: Type "ham groups in my area" into Google.
- Everyone has a website these days; this is a straightforward way to find local groups.
- Facebook: Type "Baofeng Radio" into Facebook groups and look at the hundreds that pop up. Some good ones that I'm involved in include:
- Radio Preppers
- USA Baofeng Owners
- Baofeng Outlaws
- Baofeng Owner's Club

STEP 7: PRACTICE & DRILL REGULARLY

Nothing beats hands-on experience for developing skills and confidence operating your Baofeng. I recommend:

- Take your radio when hiking and camping for actual field use.
- Practice scanning all frequencies to determine active ones in your area. Try this at different times of the day and on different days of the week. Write down the frequencies.
- Use an online resource like www.repeaterbook.com to find local repeaters and practice hitting them from different locations.

- Develop a cheat sheet of frequently used settings like squelch, power, and bandwidth optimized for your common locations.
- Drill with emergency simulations—e.g., set up a roadside assistance scenario communicating within a separate partner vehicle. Note where you can and can't hear them. Try adjusting the settings and note how that changes the connection.

Regular practical use makes quickly configuring and operating the UV-5R second nature. Applied experience provides the most significant learning.

FINAL ADVANCED STEP

After following this initial checklist, you now have an optimized foundation to enjoy your new skill. Next, continue expanding your knowledge by:

- Study for the amateur radio license exam and get your license.
- Consider getting your friends and family started in the hobby by gifting Baofeng handhelds. Pre-program their radios to your local channels. Start building your neighborhood net.

The world of amateur radio is rich, and this guide only scratches the surface. But by taking it step-by-step, you now have an optimal foundation. I wish you the best.

Jared@MorseCodePublishing.com

GLOSSARY

Amplitude Modulation (AM): Modulation technique that encodes information by varying the amplitude or strength of the radio carrier signal. Used in applications like AM broadcast radio. More susceptible to noise than FM.

Analog: Representation of signals as continuous waveforms rather than discrete digital data. Analog voice transmission is common in handheld radios.

Attenuation: refers to the decrease in signal strength as a radio wave travels through space or materials. It can be caused by distance, physical obstructions (buildings, terrain), atmospheric conditions, and frequency dependency.

Bandwidth: The range of frequencies a radio can receive or transmit on. Wider bandwidth allows receiving multiple signals simultaneously.

Base Station: A fixed radio installation, typically at a home or command center, used for effective long-range communication. It often includes a more powerful radio unit and larger antenna systems than portable units.

Bluetooth: is a wireless technology standard for exchanging data over short distances. It is increasingly used in radios for audio accessories, data transfer, and programming.

Call Sign: A unique identifier assigned to a radio operator by their country's telecommunications authority. It's used to legally identify the operator or station during transmissions.

Carrier Squelch: is a radio receiver function that mutes audio unless a signal is detected, preventing static noise. It activates with an incoming transmission's carrier wave, allowing users to hear only actual communications. This feature commonly improves the listening experience by filtering out background noise.

Channel: A labeled radio memory location containing preset receive and transmit frequencies. For example, channel 1 might include 151.625 MHz for convenient recall. Channels allow quick switching between saved frequencies.

CTCSS (Continuous Tone-Coded Squelch System): Subaudible tones are transmitted with your audio to prevent unwanted receptions from others on the same radio frequency.

DCS (Digital Code Squelch): Digitally encoded subaudible codes are transmitted with your audio to block unwanted receptions unless the code matches.

Digital: Representation of signals as discrete binary data rather than continuous waveforms. Digital modes convert voice and data to binary for transmission.

DTMF (Dual-Tone Multi-Frequency) is signaling used to communicate keypad button presses as tone codes in radios and phones. It is often used for repeater access.

Dual-Band radio: A radio that can operate on two distinct frequency bands, like VHF and UHF, allowing increased versatility.

Dual Watch: Monitoring two different frequencies or channels simultaneously by rapidly switching between them.

Duplex: Using two distinct frequencies for transmitting and receiving, allowing simultaneous sending and receiving.

DXing: is the hobby of receiving or sending long-distance communications. DX is a telegraphic shorthand for "distance" or "distant."

Elmer: An informal term used in the ham radio community to refer to a more experienced ham who mentors or guides newcomers to the hobby.

Fox Hunt: An amateur radio activity where participants use radio direction-finding techniques to locate a transmitter at an unknown location.

Frequency: The radio wave frequency that a radio transmits and receives on, measured in Hz or MHz. For example, "146.520 MHz" is a frequency used for 2-meter amateur radio. Frequencies are like radio station channel numbers.

Frequency Modulation (FM): A modulation technique that encodes information by varying the frequency of the radio carrier signal. It is used in two-way handheld radios for voice communication and provides a good signal-to-noise ratio.

Ham Radio (Amateur Radio): is a hobby involving radio frequencies for non-commercial communication, experimentation, and emergency services. It requires licensing.

High Gain Antenna: This type of antenna amplifies signal strength in a specific direction, enhancing long-distance communication. Its focused energy requires precise alignment for optimal performance. High-gain antennas are ideal for targeted signal transmission, measured in decibels (dB), and are less effective for omnidirectional reception.

Impedance: In radio, impedance refers to a circuit's resistance to the flow of alternating current (AC). It's a combination of resistance and reactance measured in ohms. Impedance is crucial in radio systems, as it matches the transmitting antenna and the receiver to ensure efficient power transfer and minimize signal reflection and loss. Proper impedance matching optimizes the performance of radio communication equipment.

IP Rating (Ingress Protection): is a standard used to define levels of sealing effectiveness of electrical enclosures against intrusion from foreign bodies (like dirt and dust) and moisture. This rating system is recognized internationally and is used to quantify the level of protection.

- **IP67 rating:** The first digit indicates the level of protection from solid objects or materials (like dust) on a scale of 0 (no protection) to 6 (dust tight). In this case, a '6' means the device is entirely dust-tight.
- The second digit represents the level of protection against liquids on a scale of 0 (no protection) to 9 (protected from prolonged immersion under high pressure, high-temperature spray downs). A '7' signifies that the device can be submerged in water up to 1 meter depth for 30 minutes without damage.

J-Pole Antenna: An omnidirectional antenna, typically for VHF/UHF bands, characterized by a distinctive 'J' shape. It's popular for its simplicity, low SWR, and no ground plane requirement.

Li-ion (Lithium-ion): The battery chemistry commonly used in modern handheld radios. Offers good energy density. Requires protection circuitry for safe charging and operation.

Mobile Radio: A radio designed for vehicle use, providing better performance than handhelds but less than base stations. They are often used in emergency vehicles and for off-road navigation.

MR (Memory Recall): Memory Mode allows users to access and operate on frequencies that have been previously programmed and stored in the radio's memory channels. When a radio is in MR (Memory Mode), you can easily switch between channels you've saved without manually entering frequencies each time.

Net Control: A designated station in an amateur radio net that manages communications and directs net operations.

NiMH (Nickel-Metal Hydride): This battery chemistry is often used in older handheld radios. It is more susceptible to performance degradation over time than Li-ion.

Offset: The difference between the receive and transmit frequencies required for repeaters. It allows simultaneous reception and transmission.

Parity: Adding a simple checksum to transmitted data bits to detect errors in received data packets. Used in digital transmission schemes.

PL Tone (Private Line): Pilot tone squelch system using subaudible tones similar to CTCSS for selectively receiving transmissions encoded with a matching tone.

Propagation: The physical means by which radio waves travel through the atmosphere and environment between transmitting and receiving antennas. It involves how signals are guided, dispersed, refracted, and absorbed. Key factors are frequency, distance, terrain, obstructions, antenna height, and atmospheric conditions. Understanding propagation allows for optimizing equipment and frequencies for the best transmission range and signal strength.

PTT (Push To Talk): The push-to-talk button initiates voice transmissions in half-duplex radio systems. The transceiver switches to transmit mode when the operator presses PTT.

Q Code: A standardized collection of three-letter message encodings, initially developed for commercial radiotelegraph communication and later adopted by other radio services, especially amateur radio.

QRP Operation: Refers to transmitting with low power. In ham radio, QRP operation is a challenge and skill, attempting to make distant contacts with minimal power output, often less than 5 watts. "QRP" does not stand for specific words; instead, it's a code derived from the Q-code system.

QSL Card: A written confirmation of two-way radio communication between two amateur radio stations or a one-way reception of a signal from an AM radio, FM radio, television, or shortwave broadcasting station. "QSL" does not represent specific words; rather, it's a code derived from the Q-code system. "QSL" originally meant "I acknowledge receipt" or "Do you acknowledge?".

Rag Chew: Informal conversation via radio, typically on non-technical subjects, often lasting for an extended period.

RDIS (Radio Data Information Service): Digital data streams encoded alongside analog FM voice transmissions to convey identifying and messaging information.

Repeater: A device that receives radio signals on one frequency and retransmits them at higher power on another frequency to extend range. For example, a repeater's input frequency might be 146.950 MHz, and the output might be 146.350 MHz.

Repeater Directory: A listing or database of repeater frequencies and information, often used by ham radio operators to find local repeaters.

Repeater Offset: The difference between the repeater's receive frequency (input) and transmit frequency (output). Allows simultaneous reception and retransmission without interference between the frequencies.

RF (Radio Frequency): RF is a fundamental concept in wireless communications, including everything from your car's radio to sophisticated satellite systems, referring to a range of electromagnetic waves that transmit data wirelessly. These waves oscillate at frequencies between 20 kHz to 300 GHz. Simply put, RF is the invisible energy that enables wireless communication.

RF Gain: Adjusts the radio's receiver sensitivity or amplification level to the strength of incoming signals.

RFA (Radio Frequency Adapter): Device that converts between different radio bands or frequency ranges. Allows interoperability between diverse radio types.

RF Interference (RFI): Unwanted reception of RF signals that disrupt normal radio operations. Various electronic devices can cause it, a common issue in densely populated areas.

RSSI (Received Signal Strength Indicator): A measurement of the strength or power level of received radio signals, displayed by the radio as a signal meter.

Simplex: A mode of communication where signals are transmitted and received on the same frequency channel but not simultaneously. In simplex, communication is unidirectional at any given time, meaning one party can either talk or listen, but not both at the same time. It's akin to a one-way street where messages travel in one direction at a time. Simplex is commonly used in basic two-way radios, such as walkie-talkies, where one user must release the push-to-talk button to listen to the other party.

SMA (SubMiniature version A): It's a type of RF (radio frequency) coaxial connector used in various applications, including antennas, radios, and telecommunications equipment. SMA connectors are known for their compact size and reliable connectivity, making them a popular choice in small devices like handheld radios, including many Baofeng models.

Squelch: Mutes the radio speaker when no transmission is received to reduce background noise. Adjusting the squelch threshold filters out weak or irrelevant signals.

SWR (Standing Wave Ratio): A measure of the radio antenna system's efficiency in transmitting power. It indicates how well the antenna matches the radio, with a lower SWR indicating better efficiency.

Tactical Call Signs: Temporary call signs are used for simplicity and clarity instead of formal call signs during emergency operations or events.

Time-Out Timer (TOT): A safety feature that limits the maximum duration of radio transmissions to prevent overheating or hogging the channel.

TPU (Thermoplastic Polyurethane) Case: A TPU case blends rubber's flexibility with plastic's strength. These cases are used for electronic devices, offering protection against impacts and scratches while being slim and resistant to oils. TPU's elasticity makes it easy to use, making it a popular choice for device protection.

UHF (Ultra High Frequency): Radio frequency ranges from 300 MHz to 3 GHz and is used for two-way radio, Wi-Fi, Bluetooth, and other applications. Baofeng radios typically operate on UHF frequencies.

VFO (Variable Frequency Oscillator): allows a radio receiver or transmitter to be tuned to different frequencies. In practical terms, when your radio is in VFO mode, you can manually enter a frequency for immediate use. This mode differs from the memory mode, where you access pre-stored and programmed frequencies.

VHF (Very High Frequency): Radio frequency ranges from 30–300 MHz, commonly used for two-way radio, FM, and television broadcasting. Some Baofeng radios also have VHF frequency capability.

VOX (Voice-Operated Transmission): The feature allows hands-free automatic transmission based on detecting your speech input via a microphone.

REFERENCES

Admin. (2023, January 3). *The best car radio antennas for 2023*. Hamtronics. https://www.hamtronics.com/best-car-radio-antennas/

Alazarsemere. (2020, May 6). *Eight maintenance tips on how to optimize the life of your radio batteries*. Tridon. https://www.tridon.com/8-maintenance-tips-to-maximize-the-life-of-your-radio-batteries/

Amateur radio service. (2016, September 28). Federal Communications Commission. https://www.fcc.gov/wireless/bureau-divisions/mobility-division/amateur-radio-service

Amateur radio terms. (2022). Boy Scouts of America. https://www.scouting.org/international/jota-joti/jota/terms/

Baofeng - Baofeng troubleshooting. (2022, January 6). *RadioReference.com* forums. https://forums.radioreference.com/threads/baofeng-troubleshooting.436574/

Baofeng cheat sheet. (n.d.). W7APK. Retrieved October 20, 2023, from https://w7apk.com/baofeng

Baofeng UV-5R+ plus review, a better UV-5R? (2022). Besthamradio. https://www.besthamradio.com/baofeng-uv-5r-plus/

Baofeng UV-82HP vs UV-5R - Which one should I buy? (2022, October 18). Walkie-Talkie-Guide. https://www.walkie-talkie-guide.com/baofeng-uv-82hp-vs-uv-5r/

Baofengradio. (2021, October 31). *Two way radio glossary of abbreviations & terms*. Baofeng. https://www.baofengradio.com/blogs/news/two-way-radio-glossary

Belousov, D. (2022, April 26). Ham radio lingo you must know. *Crunch Reviews*. https://www.crunchreviews.com/blog/ham-radio-lingo-you-must-know/#What_Are_the_Available_Types_of_Ham_Radio_Licenses

Best Baofeng radios. (2022). *Buyersguide*. https://buyersguide.org/Baofeng-radio/t/best

Best Baofeng walkie talkies. (n.d.). *Buyersguide*. https://buyersguide.org/Baofeng-walkie-talkie/t/best?Country=US&msclkid=026b d3e8294c1be58a64b772deac53b4&m=e&d=c&c=77584480440879

Buytwowayradios. (2016a, July 18). What to do if the radio is damaged. Buy two way radios. https://www.buytwowayradios.com/blog/2016/07/what_to_do_if_the_radio_is_damaged.html

Buytwowayradios. (2016b, August 1). What to do if the programming cable does not detect the radio. *Buy two way radios*. https://www.buytwowayradios.com/blog/2016/08/programming_cable_does_not_detect_the_radio.html

Buytwowayradios. (2016c, July 29). What to do if the programming cable does not work. *Buy two way radios*. https://www.buytwowayradios.com/blog/2016/07/programming_cable_does_not_work.html

Buytwowayradi. (2021) The difference between FRS and GMRS radios, Buy Two Way Radios. https://www.buytwowayradios.com/blog/2021/10

Centers, J. (2020, December 4). *Baofengs are fine radios, but they don't have hundreds of miles of range*. The Prepared. https://theprepared.com/blog/baofengs-are-fine-radios-but-they-dont-have-hundreds-of-miles-of-range/

Centers, J. (2021, March 24). *How to manually program a BaoFeng radio*. The Prepared. https://theprepared.com/gear/guides/baofeng-radio-manual-programming/

Centers, J. (2022, January 29). *How to program Ham radios with CHIRP software*. The Prepared. https://theprepared.com/gear/guides/program-ham-radio-chirp-software/

Chambers, R. (2021, August 13). *10 Best Baofeng radios (2022 update) buyer's guide*. Best Survival. https://bestsurvival.org/best-baofeng-radio/

Chen, J. (2021, November 30). The facts about distance: What's the range for radio? *Baofeng*. https://www.baofengradio.com/blogs/news/the-facts-about-distance

Chief Baofener. (2023, June 13). How to adjust squelch levels on a Baofeng UV-82 radio or crystal-clear communications. *Baofeng Academy*. https://baofengacademy.com/uv82/how-to-adjust-squelch-levels-on-a-baofeng-uv-82-radio-or-crystal-clear-communications

Chief Baofener. (2023a, April 3). Tips and tricks for using your Baofeng UV-82 radio in extreme weather conditions. *Baofeng Academy*. https://baofengacademy.com/uv82/tips-and-tricks-for-using-your-baofeng-uv-82-radio-in-extreme-weather-conditions

Chief Baofener. (2023b, January 25). Troubleshooting common issues with the Baofeng UV-82. Baofeng Academy. https://baofengacademy.com/uv82/troubleshooting-common-issues-with-the-baofeng-uv-82Baofener.

Chief Baofener. (2023b, March 28). 5 tips for maximizing battery life on your Baofeng UV-82 radio. *Baofeng Academy*. https://baofengacademy.com/uv82/5-tips-for-maximizing-battery-life-on-your-baofeng-uv-82-radio#:

Chief Baofener. (2023c, January 25). Troubleshooting common issues with the Baofeng UV-82. *Baofeng Academy*. https://baofengacademy.com/uv82/troubleshooting-common-issues-with-the-baofeng-uv-82

Chief Baofener. (2023d, April 8). 10 Off-grid survival tips for using your Baofeng UV-82 radio. *Baofeng Academy*. https://baofengacademy.com/uv82/10-off-grid-survival-tips-for-using-your-baofeng-uv-82-radio

Choosing an antenna. (n.d.). Practical Antennas. https://practicalantennas.com/about/choosing/

Contributor, S. S. (2023, June 6). *Best survival radios: Why you must secure one (before SHTF)*. Skilled Survival. https://www.skilledsurvival.com/survival-radio-communication-in-a-broken-world/

Cribbs, J. (2019a, November 23). *Ham radio emergency communications guide*. Ham Radio Prep. https://hamradioprep.com/ham-radio-in-emergencies/

Cribbs, J. (2019b, December 18). *The ultimate Baofeng guide for ham radio*. Ham Radio Prep. https://hamradioprep.com/baofeng-vhf-and-uhf-handheld-radio/

Draper, J. (2023, January 19). Best Baofeng radio in 2023 - Reviews & buying guide. *Savenetradio*. https://www.savenetradio.org/best-baofeng-radio/

Encyclopædia Britannica, Inc. (2022). *How radio works using radio waves*. Britannica. https://www.britannica.com/video/214986/How-radio-works-overview-radio-waves-frequency-amplitude-modulation

Enforcement overview. (2020). Federal Communications Commission. https://www.fcc.gov/sites/default/files/public_enforcement_overview.pdf

Family radio service (FRS). (2011, August). Federal Communications Commission. https://www.fcc.gov/wireless/bureau-divisions/mobility-division/family-radio-service-frs

4x4tographer. (2021, January 10). Electrical interference / Baofeng fix. *Offroad Passport*. https://offroadpassport.com/forums/topic/5718-electrical-interference-baofeng-fix/

Frank, A. (2023, October 2). Why is my radio static all of a sudden and how to fix it. *Windupradio*. https://windupradio.com/why-is-my-radio-static-all-of-a-sudden/

General mobile radio service (GMRS). (n.d.) Federal Communications Commission. https://www.fcc.gov/wireless/bureau-divisions/mobility-division/general-mobile-radio-service-gmrs

General mobile radio service. (2023, September 6). Wikipedia. https://en.wikipedia.org/wiki/General_Mobile_Radio_Service

GMRS. (n.d.) *Radio Made Easy*. https://radiomadeeasy.com/gmrs-guide/

Grandy, G. (2022). *BaoFeng BF-F8HP review*. GearLab. https://www.outdoorgearlab.com/reviews/camping-and-hiking/walkie-talkies/baofeng-bf-f8hp#:

Ham radio licenses. (n.d.). *ARRL*. Retrieved October 20, 2023, from https://www.arrl.org/ham-radio-licenses

Ho, J. (2023, October). 5 Best car radio antennas - Oct. 2023. *BestReviews*. Retrieved October 20, 2023, from https://bestreviews.com/automotive/stereos/best-car-radio-antennas

How far can I talk? (n.d.). Quality2wayradios. https://quality2wayradios.com/store/radio-range-distance

How long will a Baofeng UV-5R portable radio last? (2019) Quora. https://www.quora.com/How-long-will-a-Baofeng-UV-5R-portable-radio-last

How to use Baofeng radios: A quickstart guide to the keypad and your programming file. (2022). Aganz Store. https://www.aganzstore.com/blog/baofeng-radios-a-quickstart-guide-to-your-programming-file

J, K. (2021, November 16). Chirp software for programming a Baofeng radio - How to use. *Modern Survival Blog*. https://modernsurvivalblog.com/communications/program-a-baofeng-radio-with-chirp-quick-start/

J, K. (2022, December 30). Baofeng extended battery for longer life between charges. *Modern Survival Blog*. https://modernsurvivalblog.com/communications/baofeng-extended-battery/

Joe. (2020, May 5). How to get your ham radio license in 3 simple steps. *Ham Radio Prep*. https://hamradioprep.com/how-to-get-your-ham-radio-license-made-easy/

Joe. (2021, April 16). Ham radio licenses: Which is right for you? *Ham Radio Prep*. https://hamradioprep.com/ham-radio-licenses/

Karonis, G. (2023a, April 12). 8 Tips to extend push-to-talk two way radio battery life. *Peak PTT Blog.* https://blog.peakptt.com/8-tips-to-extend-push-to-talk-two-way-radio-battery-life/

Karonis, G. (2023b, May 24). Clear communication tips on push to talk two way radios. *Peak PTT Blog.* https://blog.peakptt.com/clear-communication-tips-on-push-to-talk-two-way-radios/

KB6NU. (2018, September 27). Did the FCC just make Baofengs illegal? Short answer: NO! - *KB6NU's Ham Radio Blog.* https://www.kb6nu.com/did-the-fcc-just-make-baofengs-illegal-short-answer-no/

Kevin. (2021, January 11). 5 Best antenna for BaoFeng in 2022 : (Reviews & buying guide). *Antenna Tactical.* https://www.antennatactical.com/best-antenna-for-baofeng/

Knoji Staff. (n.d.). How long does BaoFeng's UV-82HP two-way radio last between charges? *Knoji.* Retrieved October 20, 2023, from https://baofeng.knoji.com/questions/baofeng-uv-82hp-two-way-radio-battery-life/

KØNR, B. (2014, February 8). Solving the Baofeng cable problem. *Amateur Radio.* https://www.amateurradio.com/solving-the-baofeng-cable-problem/

Martens, M. (2014, November 15). Get a better signal out of your handheld radio. *KB9VBR Antennas.* https://www.jpole-antenna.com/2014/11/15/get-a-better-signal-out-of-your-handheld-radio/

Matt. (2023, January 24). Resource review: The guerilla's guide to the Baofeng radio. *Everyday Marksman.* https://www.everydaymarksman.co/resources/guerillas-guide-to-the-baofeng/

Mclaren, S. (2023, June 23). Ham radio emergency frequencies and common uses [PDF]. *Stryker Radios.* https://strykerradios.com/ham-radios/ham-radio-emergency-frequencies-common-uses/#:

McNany, A. (2016, February 18). Baofeng UV-5R: the classic Chinese handheld. *ARTech.* https://amateurradiotech.com/baofeng-uv-5r/

Meacher, N. (2019, August 12). New FCC rules that impact preppers and new hams. *Survival Dispatch.* https://survivaldispatch.com/new-fcc-rules-that-impact-preppers-and-new-hams/

Miklor Home. (n.d.). Common error messages. *Miklor.* https://www.miklor.com/COM/UV_ErrorMess.php

Noonan, T. (n.d.). BaoFeng UV-5R - Emergency communication & frequencies. *Bug out Bag Builder.* https://www.bugoutbagbuilder.com/learning-tutorials/baofeng-uv-5r-basic-setup-emergency-frequencies

Overlandbound. (2016, November 16). OB approved - Baofeng radio configuration and methods. OVERLAND BOUND COMMUNITY. https://www.overlandbound.com/forums/threads/baofeng-radio-configuration-and-methods.3330/

Pekelny, J. (2021, July 14). Review: Yaesu FT-60R vs BaoFeng BF-F8HP for new ham radio operators. *The Prepared.* https://theprepared.com/blog/review-yaesu-ft-60r-vs-baofeng-bf-f8hp-for-new-ham-radio-operators/

Preparation Quotes. Notable Quotes. Accessed January 18, 2024. https://www.notable-quotes.com/p/preparation_quotes.html.

Problem with Boafenf BF-888 radios. (n.d.). *Radio Stack Exchange*. Retrieved October 20, 2023, from https://ham.stackexchange.com/questions/17596/problem-with-baofeng-bf-888-radios

Quinton. (2023, June 2). Baofeng BF-F8HP vs UV-82HP- Ham radio planet. *Ham Radio Planet*. https://hamradioplanet.com/baofeng-bf-f8hp-vs-uv-82hp/

Radio communications for survivalists and preppers. (n.d.). *Bestthamradio*. Retrieved October 20, 2023, from https://www.besthamradio.com/radio-communications-for-survivalists-and-preppers/

Radio frequencies you need to know for emergencies. (2023, January 5). 4Patriots. https://4patriots.com/blogs/news/radio-frequencies-you-need-to-know-for-emergencies

Radioreference. (2013, June 7). How to program repeaters into a Baofeng UV-5R. RadioReference.com forums. http://forums.radioreference.com/threads/how-to-program-repeaters-into-a-baofeng-uv-5r.268176/

RatedRadarDetector. (2023, October 3). What is the best Baofeng radio?. https://www.ratedradardetector.org/best-baofeng-radio/

Reed, A. (2020, March 8). 7 best FM antennas of 2020 to have a crystal clear reception - AW2K. *A Web Not to Miss*. https://www.awebtoknow.com/reviews/best-fm-antennas/

Retevis. (2017, February 19). *Squelch and squelch level*. Two Way Radio Community. https://www.twowayradiocommunity.com/squelch-squelch-level/

Richard. (2020a, June 7). 7 Best Baofeng radios of 2023 - Reviews & buying guides. *Amateur Radio Wiki*. https://www.amateur-radio-wiki.net/best-baofeng-radio/

Richard. (2020b, September 18). 7 Best AM/FM antennas 2023 - Indoor & outdoor radio antenna reviews. *Amateur Radio Wiki*. https://www.amateur-radio-wiki.net/best-fm-radio-antenna/

Rick. (2013, January 21). Radio 101 - Using VOX on two way radios. *Buy Two Way Radios*. https://www.buytwowayradios.com/blog/2013/01/radio_101_-_vox.html

Rick. (2016, July 27). What to do if the radio programming software does not detect the cable or COM port. *Buy Two Way Radios*. https://www.buytwowayradios.com/blog/2016/07/programming_software_does_not_detect_cable_or_com_port.html

Rick. (2016, July 6). What to do if the radio does not power on. *Buy Two Way Radios*. https://www.buytwowayradios.com/blog/2016/07/what_to_do_if_the_radio_does_not_power_on.html

Ricky. (2022, January 11). How to choosing the right radio antenna for your car. Sound Audio Guru. https://soundaudioguru.com/chossing-radio-antenna-for-car#:

Rock Networks (n.d.-a). *10 Steps to extending battery life in two way radios*. Retrieved October 20, 2023, from https://www.rocknetworks.com/lengthening-portable-two-way-radio-batteries-with-proper-maintenance

Rock Networks. (n.d.-b). *How to select the right frequency for your two way radio.* . Retrieved October 20, 2023, from https://www.rocknetworks.com/how-to-select-the-right-two-way-radio-frequency/

Sampson, M. (2022, June 24). The best long-range walkie talkies for clear comms. *Task & Purpose*. https://taskandpurpose.com/gear/best-long-range-walkie-talkies/

SDR, O. (2021, August 21). Best antenna for Baofeng radio in 2023 - Performance upgrades. *OneSDR - a Wireless Technology Blog*. https://www.onesdr.com/best-antenna-for-baofeng-radio/

Sher, M. (2022, November 4). How to take care of two-way radios. *Hytera*https://www.hytera.us/resources/how-to-take-care-of-two-way-radios/

Six tips to maintain your two way radio. (2014). *Two Way Radio Gear*. https://twowayradiogear.com/blogs/news/six-tips-to-maintain-your-two-way-radio

Smith, D. (n.d.). Beginners guide - CHIRP. *Chirp.danplanet*. https://chirp.danplanet.com/projects/chirp/wiki/Beginners_Guide

Storage of a Baofeng UV-5R. (n.d.). *Amateur Radio Stack Exchange*. Retrieved October 20, 2023, from https://ham.stackexchange.com/questions/3488/storage-of-a-baofeng-uv-5r

Superior Support. (2022). *Model type of radio primary user frequency range additional options PTT style speaker size display*. Baofeng Tech. https://baofengtech.com/wp-content/uploads/2020/10/CompareChart.pdf

TalkieMan. (2019a, June 13). How far can a Baofeng transmit? *Talkie Man*. https://talkieman.com/how-far-can-a-baofeng-transmit/

TalkieMan. (2019b, September 14). Baofeng BF-888S review. *Talkie Man*. https://talkieman.com/baofeng-bf-888s-review/

Tate, A. (2021, August 28). Real life survival story: Baofeng, the little ham radio that could -. *The Organic Prepper*. https://www.theorganicprepper.com/survival-story-baofeng-ham-radio/

Thomas, T. (2022). How to clean and care for your two-way radio (+ 2 printable lists). *Waveband Communications*. https://www.wvbandcoms.com/blogs/blog/cleaning-two-way-radio

Tobias. (2013, September 1). Long distance communications with ham radios & satellites. *GeekPrepper*. https://geekprepper.com/long-range-communications-using-ham-radios-satellites/

Tobias. (2015, February 22). How program a Baofeng UV-5R with CHIRP software. *GeekPrepper*. https://geekprepper.com/program-a-baofeng-radio-with-chirp/

Twowaydirect. (n.d.). How to preserve your two way radios. *Two Way Direct*. https://www.twowaydirect.com/two-way-direct-blog/how-to-preserve-your-two-way-radios/

UV-5R high gain antenna performance. (n.d.). *Amateur Radio Stack Exchange*. Retrieved October 20, 2023, from https://ham.stackexchange.com/questions/18716/uv-5r-high-gain-antenna-performance

Vivian. (2021, May 25). What is the difference between CTCSS and DCS on two-way radios? *Ailunce* blog. https://www.ailunce.com/blog/What-is-the-difference-between-CTCSS-and-DCS-on-two-way-radios

W0SJF. (n.d.). *Forum» General technical Q & A» Battery options for Baofeng*. ARRL. Retrieved October 20, 2023, from https://www.arrl.org/forum/topics/view/1764

Walkie-talkie-guide. (2022, May 10). Tutorial: Baofeng UV-5R programming. *Www.walkie-Talkie-Guide.com*. https://www.walkie-talkie-guide.com/baofeng-uv-5r-programming/